Mathematical Biosciences Institute Lecture Series

The Mathematical Biosciences Institute (MBI) fosters innovation in the application of mathematical, statistical and computational methods in the resolution of significant problems in the biosciences, and encourages the development of new areas in the mathematical sciences motivated by important questions in the biosciences. To accomplish this mission, MBI holds many week-long research workshops each year, trains postdoctoral fellows, and sponsors a variety of educational programs.

The MBI lecture series are readable, up to date collections of authored volumes that are tutorial in nature and are inspired by annual programs at the MBI. The purpose is to provide curricular materials that illustrate the applications of the mathematical sciences to the life sciences. The collections are organized as independent volumes, each one suitable for use as a (two-week) module in standard graduate courses in the mathematical sciences and written in a style accessible to researchers, professionals, and graduate students in the mathematical and biological sciences. The MBI lectures can also serve as an introduction for researchers to recent and emerging subject areas in the mathematical biosciences.

Marty Golubitsky, Michael Reed
Mathematical Biosciences Institute

More information about this series at http://www.springer.com/series/13083

Mathematical Biosciences Institute Lecture Series
Volume 1: Stochastics in Biological Systems

Stochasticity is fundamental to biological systems. While in many situations the system can be viewed as a large number of similar agents interacting in a homogeneously mixing environment so the dynamics are captured well by ordinary differential equations or other deterministic models. In many more situations, the system can be driven by a small number of agents or strongly influenced by an environment fluctuating in space or time. Stochastic fluctuations are critical in the initial stages of an epidemic; a small number of molecules may determine the direction of cellular processes; changing climate may alter the balance among competing populations. Spatial models may be required when agents are distributed in space and interactions between agents form a network. Systems evolve to become more robust or co-evolve in response to competitive or host-pathogen interactions. Consequently, models must allow agents to change and interact in complex ways. Stochasticity increases the complexity of models in some ways, but may smooth and simplify in others.

Volume 1 provides a series of lectures by well-known international researchers based on the year on Stochastics in Biological Systems which took place at the MBI in 2011–2012.

Michael Reed, Richard Durrett
Editors

Mathematical Biosciences Institute Lecture Series
Volume 1: Stochastics in Biological Systems

Stochastic Population and Epidemic Models
Linda S. Allen

Stochastic Analysis of Biochemical Systems
David Anderson; Thomas G. Kurtz

Stochastic Models for Structured Populations
Vincent Bansaye; Sylvie Méléard

Branching Process Models of Cancer
Richard Durrett

Stochastic Neuron Modeling
Priscilla E. Greenwood; Lawrence M. Ward

The Mathematics of Intracellular Transport
Scott McKinley; Peter Kramer

Population Models with Interaction
Etienne Pardoux

Correlations from Coupled Enzymatic Processing
Ruth Williams

Priscilla E. Greenwood • Lawrence M. Ward

Stochastic Neuron Models

Springer

Priscilla E. Greenwood
University of British Columbia
Vancouver, Canada

Lawrence M. Ward
University of British Columbia
Vancouver, Canada

ISSN 2364-2297   ISSN 2364-2300   (electronic)
Mathematical Biosciences Institute Lecture Series
ISBN 978-3-319-26909-2   ISBN 978-3-319-26911-5   (eBook)
DOI 10.1007/978-3-319-26911-5

Library of Congress Control Number: 2015960199

Mathematics Subject Classification (2010): 35Q84, 37H10, 39A21, 60I20, 60H30, 60I70, 62P10, 92C20

Springer Cham Heidelberg New York Dordrecht London

Springer International Publishing AG Switzerland is part of Springer Science+Business Media (www.
springer.com)

Preface

In this book we describe a large number of open problems in the theory of stochastic neural systems, with the aim of enticing probabilists to work on them. These include problems arising from stochastic models of individual neurons as well as those arising from stochastic models of the activities of small and large networks of interconnected neurons. We sketch the necessary neuroscience background to these problems so that probabilists can grasp the context in which they arise. We focus on the mathematics of the models and theories, rather than on descriptions of empirical systems. Although considerable insight into the behavior of these models has been developed through computer simulation, mathematical results are, at present, limited. We hope to inspire a probabilistic attack on these open problems that will advance theoretical neuroscience.

Several themes appear herein. A stochastic model of a dynamical system is often formulated first in terms of local Poisson rates, with a discrete state space of counts. A large N approximation, which recently has been termed the "linear noise approximation," and can be thought of as a stochastic process version of a linear approximation, appears in a number of contexts here.

Another theme is oscillations sustained by noise. If a deterministic model has oscillations that are damped to a fixed point, a stochastic version of the model will have sustained, stochastic, oscillations called "quasicycles." The phenomenon appears in the context of a single neuron where stochastically sustained oscillations provide a useful model of subthreshold oscillations. It appears at the population level where stochastically sustained oscillations provide an explanation of observed population gamma-band rhythms. The same phenomenon also appears spatially, where patterns that would be damped in a deterministic model are sustained by stochasticity and are called "quasipatterns."

Our mathematical focus is on stochastic dynamical systems defined by small sets of interacting stochastic differential equations. The text heavily emphasizes work done by the authors together with several collaborators. We thank them for their important contributions to that work. We are also grateful for support from the

Natural Sciences and Engineering Research Council of Canada and from the Peter Wall Institute for Advanced Studies at the University of British Columbia, which was instrumental in initiating several of our collaborations.

Vancouver, BC, Canada   Priscilla E. Greenwood
                                                                  Lawrence M. Ward

Contents

Chapter 1
Introduction

Neuroscience has become a vast field. Within it the modelling of neural systems is a small corner. Within that small corner the portion attending to stochastic effects is again small. Nevertheless it is a large topic, and we will tell you about only a subset that we happen to know something about. A lot of basic work has been done by researchers with limited background in probability, and simulation, as a method, is far ahead of stochastic analysis. The result is a field rich in opportunity for probabilists. We will tell you about constructions and results, trying to supply details to the extent necessary to get you started thinking about problems. These problems will be labeled with the symbol $\mathcal{PP}x.y.z$, where x is the chapter number, y is the section number within that chapter, and z is the problem number within that section. They will be set off as separate paragraphs from the rest of the text.

This introduction will continue, in Section 1.1, by pointing out where noise, i.e. stochasticity, arises in the functioning of neurons and neural systems. Of course probabilists believe that everything is stochastic. Section 1.2 reviews the topic of *stochastic facilitation*, the term we use, following [1], for what has confusingly been called 'stochastic resonance.' Our view is that nearly all neural processes operate by making use of stochastic effects, and would come to a sad halt without them. We also find, however, that understanding of neural processes begins with capturing an underlying deterministic dynamical system at work. We will, in fact, refer to a few basic ideas from dynamical systems theory without much fanfare.

A proper study of this topic should include much about neurobiology. There are many excellent texts on this topic, but the point of this work is to identify problems on which probabilists can make progress without studying those texts. Once you get hooked on a problem you can study the relevant neurobiology.

In Chapter 2 we describe several models of single neurons. Formally, a neuron has an input, internal dynamics, and an output, all functions of discrete or continuous time. For a real neuron, the input is through synapses onto the neuron's dendrites, or onto the neuron's cell body, its soma. The dendrites are fine projections from a stalk that projects out of the soma, and synapses chemically connect sites on the dendrites

© Springer International Publishing Switzerland 2016
P.E. Greenwood, L.M. Ward, *Stochastic Neuron Models*, Mathematical Biosciences
Institute Lecture Series, DOI 10.1007/978-3-319-26911-5_1

or the soma to axons (like little cables) from other neurons. The internal dynamics take place inside the soma, and the output is through axons that connect chemically to other neurons.

A formal, or model, neuron is described in terms of its 'state' at each time, t. The state is a set of real numbers giving the output and values of internal dynamics at time t. The output can be coded as '0' or '1.' When the output at time t is '1' the neuron is said to have 'fired' at time t. Firing is equivalent in a neuron model to what is called in neurobiology an 'action potential' or 'spike' generated by a neuron. Single neuron models fall into three classes: binary neurons, those having only two states; integrate and fire neurons, where subthreshold dynamics are modelled but firing is treated as an instantaneous event; dynamical systems models, where a set of continuous variables describes firing and subthreshold dynamics in one package. It is worthwhile to understand something about single neuron models and associated problems before going on to consider models of populations of neuron models. In what follows we will often omit the word 'model' when describing neuron models and models of populations of neurons, referring instead to 'neuron(s)' or 'models of populations of neurons.' But in fact we will always be referring to neuron models and not to real neurons.

During the discussion of single neuron models the following questions should come to mind: Which model is most accurate? Which is best in some defensible sense? Does it matter, for our understanding of neural systems, how single neurons are modelled? Some indications of answers to these questions begin to appear in later sections, primarily via simulations. Most questions that occur to us are open. Probabilists will suspect that answers in the form of asymptotic equivalences should be attainable. For example, we suspect that some distributional behaviors of a population of neurons, studied in Chapter 3, are the same no matter whether the single neurons in the population are modelled by an integrate-and-fire model or by a more elaborate dynamical systems model. Numbers of neurons are large in many neural systems, but still finite, and the effects of noise in the nervous system are often thought to be important because of finite size effects.

Once we have a stochastic neuron model that is a Markov process, a Kolmogorov equation can be written. If we think of a population of neurons, or a population of excitatory–inhibitory neuron pairs (see Chapter 3) as governed by i.i.d. Markov processes, then the Kolmogorov equation for the single neuron gives the evolution of the probability distribution of the family of processes on its state space. Recent advances allow computation of the solution of complicated Kolmogorov equations, giving response to a functional input in real time [2]. Although the Kolmogorov equation contains all of the distributional information about the process, the sample-path viewpoint generally provides easier access to insights about process behavior. In addition it is awkward to incorporate dependencies starting with a distributional viewpoint. Here we will concentrate instead on a stochastic dynamics approach, where a neuron or population of neurons is regarded as a dynamical system, satisfying a family of ODEs, or PDEs in spatial models, that become SDEs, or SPDEs, when stochastic effects are incorporated.

As we turn, in Chapter 3, to models of populations of neurons, an important question arises: What variable is convenient as a measure of population activity? We make a case for using a summary voltage of the population, which may covary with the firing rate, for this purpose. Each neuron is either excitatory or inhibitory, meaning that it excites or inhibits other neurons on which it impinges. Excitatory and inhibitory neurons tend to be connected in pairs. In general, population models are defined in terms of excitatory and/or inhibitory neurons and their rates of firing.

We pay particular attention to a Wilson-Cowan type model for the voltages of interacting populations of excitatory and inhibitory neurons, which damp to a fixed point [3]. The sustained oscillations present in the stochastic model are subjected to stochastic analysis that reveals its phase and radial processes in an explicit form allowing further computation. This example raises optimism for wider application of stochastic analysis here.

The different approach of Brunel and Hakim [4] to neuronal population modelling involves explicit delay and families consisting entirely of inhibitory neurons.

$\mathcal{PP}$1.0.1 We suggest in Section 3.4 that the Wilson-Cowan and Brunel-Hakim approaches might be reconciled via stochastic analysis.

We finish the chapter on population models with a simulation result that indicates there is a critical value of Kuramoto-type coupling strength at which quasicycles are abruptly synchronized when populations are sufficiently large.

Spatially structured neural systems are beginning to receive attention from stochastic neuronal modelers. Chapter 4 proposes that stochastic reaction–diffusion equations can be used to describe developing and sustained cortical patterns that may result from chemical imbalances, or that may exist in untrained or unengaged visual cortex. Stochastic Turing patterns appear in Chapter 4 as spacetime analogues to the sustained oscillations described in Chapter 3.

Late in Chapter 4 we gather additional thoughts about spatial, and, more generally, graphical neural models. Various graphical structures have been explored via simulation, both regarding synchronization and regarding formation of large subgraphs. We expect that the stochastic community studying graphical structures will find both problems and applications of existing results in this direction. We finish in Chapter 5 with a sketch of some 'big picture' brain simulations.

1.1 Sources and modelling of neuronal stochasticity

At the outset it is useful to have an overview of the sources and modelling of neuronal noise. The review that inspired this section was that of Longtin [5]. Several other recent discussions of neural noise also could be consulted (e.g., see [1] and the references therein).

A typical neuron responds differently upon each presentation of a repeated signal. In fact a neuron in its normal environment among other neurons is constantly subject to all sorts of inputs, which are often irrelevant to its momentary computational purpose. A neuron subject to irregular input of this kind, or even to 'constant' input, fires irregularly if it is, e.g., a pyramidal neuron (the most common type of neuron in the neocortex with a soma shaped like a pyramid), or, in contrast, fires regularly if it is a pacemaker neuron (a type of neuron whose function is to provide a kind of clock-like timing signal).

Firing variability is often seen as a manifestation of neural noise. Noise occurs at many scales. Thermal noise, which occurs in every system, is usually considered to be too weak to be relevant in neural systems. Its presence, however, underlies the motions of ions through ion channels in the cell walls of a neuron. The opening and closing of ion channels is considered to be one of the main sources of intrinsic neuronal noise. Nanoscale modelling of how ion channels open and close in response to electrical and chemical changes [6, 7] has been used in computation of interspike interval distributions [8].

Identifying noise sources may be challenging. For instance, in a stochastic dynamic model of a neuron, the effects of noise arising from synaptic input or from ion channel fluctuations may have virtually identical effects on the response of the model to an assortment of other inputs.

$\mathcal{PP}$1.1.1 This problem could be an engaging one for probabilists. It will arise in the study of the Izhikevich neuron in Section 2.3 or any of the other multidimensional SDE models of Chapter 2.

In fact, synaptic noise, that is, irregular inputs from other neurons, has been considered to be the dominant noise source when we work at the level of neural networks. Here we are thinking of data from in vivo recordings, where noise increases the mean conductance of a neuron, compared to in vitro recordings, where synaptic stimulation is weak.

Some deterministic single cell models produce chaos if driven by a constant input [9]. It has been said that when fluctuations of synaptic input yield a strongly irregular pattern of activity, we have a signature of deterministic chaos [10]. We know, however, that it is statistically impossible to separate mathematical chaos from stochasticity in general (e.g., [11]). Because noise is everywhere, and we are probabilists, we confine ourselves here to stochastic models.

1.2 Stochastic facilitation

As a general rule, noise in a system is considered to be bad. It should be minimized, or better, eliminated, in order for the system to operate optimally. We are convinced, however, that in the nervous system this is not the case. Too much noise, certainly, could be detrimental to the functioning of neurons, or systems of neurons. But a system of neurons, according to much data and much modelling, does not function with no stochasticity at all [1]. Hence, there is an optimal level, or possibly several optimal levels (variances), of noise that produces the best functioning.

The idea that there is an optimal level of noise to augment information transmission has had great popularity in several areas, going under the name 'stochastic resonance.' The name began to be used in contexts where a periodic motion came into 'stochastic resonance' with a random process at a certain level of stochasticity, and then spread to many other contexts where the name was somewhat inappropriate (there being no resonance per se), and thus caused some confusion. We will avoid using this term and instead call a situation where a noise helps a neural system to function *stochastic facilitation* [1].

A question for probabilists is of the form: in a particular model find the level of noise such that information in the output regarding the input is optimized. The measure of the information in the input that can be found in the output has sometimes been signal-to-noise ratio, correlation, or as recommended in [12, 13], Fisher Information. We will point out several such problems, where the measure of information transfer is appropriate to functional, or computational, roles of neural systems.

An early paper that helped popularize the idea that neuronal functioning could be best served by an optimal level of noise, at least in the sense of its output reflecting a subthreshold periodic input, was [14]. In that paper, Longtin shows that a stochastic bistable model, forced with a sinusoid,

$$\frac{dx}{dt} = -\frac{dU(x)}{dx} + \eta(t) + m\,sin(\omega t), \tag{1.1}$$

produces an interspike interval (ISI) histogram as in Figure 1.1, from which one can read off the forcing frequency, ω. Here $U(x)$ is a double-well potential, namely $U(x) = x - x^3$, $\eta(t)$ is an Ornstein-Uhlenbeck process satisfying

$$d\eta = -\lambda\,\eta(t)dt + \lambda\,d\xi(t), \tag{1.2}$$

and $\xi(t)$ is a standard Brownian motion. The model does not really spike, of course, but rather alternates between the two potential wells. An 'interspike interval' is defined as the time it takes after leaving, say, the first well, to return there. In Figure 1.1 we see from the histogram that, roughly, if the transition is not made in one cycle of the forcing sinusoid, then it is likely to be made in two cycles, and if not in two then in three, and so on.

We will see that a stochastic dynamical neuron firing model typically has an internally-generated frequency arising from its deterministic part. This, together with the noise, produces an effect in the ISI histogram similar to that in Figure 1.1. An example for the Hodgkin-Huxley neuron is Figure 1.2. We write more about this in Section 2.4.

$\mathcal{PP}$1.2.1 A problem for probabilists is to use stochastic analysis, or simulation, to determine whether the natural level of noise, based on estimated parameters, is optimal from the point of view of stochastic facilitation in various settings.

$\mathcal{PP}$1.2.2 A second problem is to clarify the common role of stochastic facilitation in Figure 1.1 and in the ISI histogram of the HH neuron appearing in Figure 1.2.

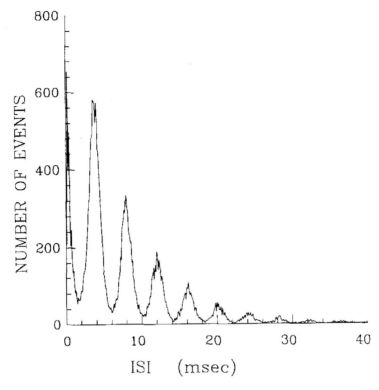

Fig. 1.1 ISI histogram for a double-well potential forced with a sinusoid and Ornstein-Uhlenbeck noise. The intervals between the peaks correspond to the period of the forcing sinusoid. Reprinted with permission from [14]

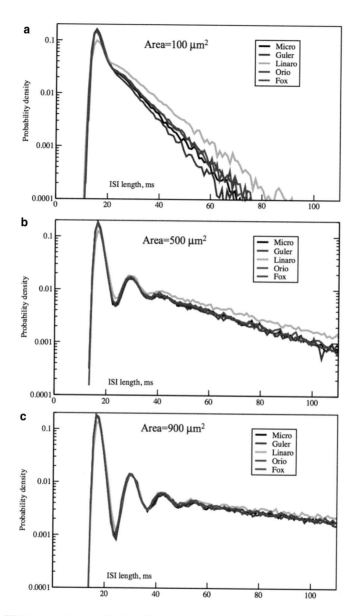

Fig. 1.2 ISI histogram for the Hodgkin-Huxley neuron computed by various algorithms (colors; explained in Chapter 2). The initial peak shows the probability density of the time interval taken by a single spike. The subsequent intervals between local minima correspond to the times taken for the sample path to circuit the fixed point of the deterministic model, and so corresponds to a forcing period as in Figure 1.1. The number of ion channels increases, and the standard deviation of the noise decreases, with increasing membrane area. A comparison with the stochastic dynamics of Figure 1.1 gives rise to a challenging problem (see text). Reprinted with permission from [8]

Chapter 2
Single Neuron Models

Neuron models seem to come in three types: binary, threshold, and dynamical. We indicate some of the potential problems for probabilists associated with each type. In each case we first describe the deterministic model and its characteristics and then indicate how introducing noise, or stochasticity, into the model affects its behavior.

2.1 Binary neurons

In 1943 McCulloch and Pitts [15] introduced a simple binary model neuron that has been used in a variety of contexts, for example as the computational element in some neural network models. This model is an extreme simplification of a neuron, and thus is of little interest to neuroscientists on its own, although neural networks built from it do display interesting behavior. The simplification consists of collapsing all of the complicated electro-chemical dynamics of neurons into two states, firing and not firing. In our descriptions of more complicated single neuron models in the following sections, we will describe these dynamics in more detail.

The McCullough-Pitts neuron model consists of a weighted sum of inputs, I_i, and a single binary output, y (Figure 2.1). The weights W_i are scaled between -1 and 1, with negative weights associated with inhibitory inputs and positive weights associated with excitatory inputs. An arbitrary threshold is designated at which the neuron 'fires' a spike, or a '1' output. The model proceeds in discrete time steps. Inputs from all sources are assumed to occur at a discrete moment in time, and a 0 or 1 is recorded, as in equations (2.1) and (2.2). The system is cycled indefinitely. The model can be adapted easily to sum inputs occurring over continuous or discrete time, although this is typically not done for this simple model. If inputs do sum over

© Springer International Publishing Switzerland 2016
P.E. Greenwood, L.M. Ward, *Stochastic Neuron Models*, Mathematical Biosciences
Institute Lecture Series, DOI 10.1007/978-3-319-26911-5_2

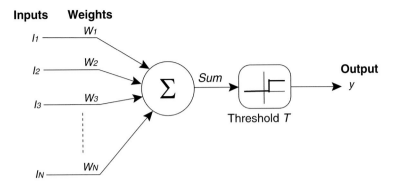

Fig. 2.1 The McCulloch-Pitts model neuron

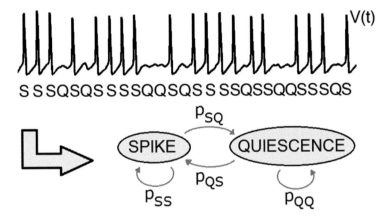

Fig. 2.2 Cycles of a binary model neuron. The illustration of $V(t)$ from a more sophisticated model shows the rough correspondence between a binary sequence and a pattern of firing and quiescence. Reprinted with permission from [16]

time, the model becomes equivalent to an 'integrate-and-fire' model without a leak (see Section 2.2). The equations defining the output at each time step are

$$V = \sum_{i=1}^{N} I_i W_i \qquad (2.1)$$

$$y = f(V) = \begin{cases} 1 \ if \ V \geq \ threshold \\ 0 \ if \ V < \ threshold. \end{cases} \qquad (2.2)$$

The model is stochastic if the inputs, I_i, include i.i.d. or correlated stochastic noise. A different binary neuron model is illustrated in Figure 2.2, where stochastic rates define the transition probabilities between the two states, to produce a simple

Markov chain with the two states called 'firing' and 'quiescent.' In that model, as in all binary models, further details of the biological neuron are assumed to be unimportant.

$\mathcal{PP}$2.1.1 An instructive exercise is to compute the distribution of outputs of the McCullough-Pitts binary neuron model described by Equations (2.1) and (2.2) as a function of stochastic input with constant weights. For example, the stochastic input could be, for each sequence of I_is, an i.i.d sequence of 0s and 1s. This exercise could be done for the output Y at a fixed time as a function of N, T, and the parameter $p(1)$ of the binary random variables.

$\mathcal{PP}$2.1.2 It could also be done for successive outputs over time, again varying N and $p(1)$.

$\mathcal{PP}$2.1.3 Also instructive would be to compute the interspike interval distribution of a sequence of outputs, where the '1' state is identified with a 'spike.'

$\mathcal{PP}$2.1.4 Another question to ask about the relationship between inputs and outputs is how much information about the inputs is preserved by the binary neuron model's firing sequence. The question of how much information can be had from a binary process is explored in [16].

2.2 Leaky integrate-and-fire neurons

The (leaky) integrate-and-fire (LIF) neuron model is by far the most widely studied and applied model of the firing (spiking) neuron. It is not just one but a large family of models. We look at a standard continuous time version. Details and a review are in [17]. The idea is to describe how the voltage across the cell membrane changes as a function of time before it crosses a threshold, at which point the neuron is deemed to generate an action potential (fire a spike), the voltage is reset to an initial value, and the process repeats. Clearly the successive firing times are equal for a deterministic model with constant input, or i.i.d. for a stochastic model with noise from a fixed probability distribution.

In a deterministic model the voltage rises or falls according to a synaptic rate, arising from input that is excitatory or inhibitory. In addition, the voltage falls in proportion to its value above a base value, V_0; this term is the 'leak.' The voltage equation takes the form

$$\tau dV_t/dt = -\lambda(V_t - V_0) + a_E S_E(t) - a_I S_I(t) \quad 0 < t < T, \quad (2.3)$$

where $T = min\{t : V_t \geq V_{threshold}\}$, $V_T = V_0$, and the process repeats. Here $S_E(t), S_I(t)$ are rates of excitatory and inhibitory synaptic input. If S_E, S_I are constants, this linear equation is easily solved.

In the stochastic model, the deterministic rates S_E, S_I are replaced by Poisson point processes with these rates, e.g., $S_E(t) = \sum \delta(t - t_{E_k})$, where the t_{E_k} are Poisson times of synaptic excitatory input of a voltage pulse. Similarly for inhibitory input, $S_I(t) = \sum \delta(t - t_{I_k})$. A Markov process is described in terms of the transition rates together with the resetting of the voltage, V_t, at each threshold crossing. The Kurtz approximation [18], which we discuss in Section 3.2, allows us to approximate such a normalized Poisson-type Markov process, when the synaptic input times become somewhat dense, by a diffusion process of the form

$$\tau dV(t) = -((V(t) - V_0) - \mu)dt + \sigma\sqrt{2\tau}dW_t, \qquad (2.4)$$

where W_t is a standard Brownian motion together with voltage resetting. The process (2.4) is an Ornstein-Uhlenbeck (O-U) process. The computation of the distribution of the time between boundary crossings, or interspike intervals (ISIs) of the stochastic LIF neuron has been of primary interest [17].

The assumption that synaptic inputs arrive at Poisson times might seem to be misleading if one looks very locally in time and if the inputs were coming from very few other neurons. In fact, however, the inputs come from many sources that are roughly independent and are observed over time intervals that contain inputs from many of the available sources. The result is that the Poisson assumption is quite good and almost universally used.

An input signal to the system may come as a time dependency of the Poisson intensity, or as a non-homogeneity in the input sources, e.g., one source becomes dominant.

The predominant focus has been on how (LIF) neurons work when synaptic input is 'constant,' meaning for us here, stochastically stationary. There is also the possibility that synaptic input is not Markov but is better represented as driven by fractional Gaussian noise [19]. A survey comparing various stochastic LIF models can be found in Chapter 5 of [17].

$\mathcal{PP}$2.2.1 The modelling of input signals and how neurons extract meaning, or information, from them is an area deserving attention from probabilists. For example, one could assess what effects the Poisson assumption has on the ISI distribution of the LIF model.

$\mathcal{PP}$2.2.2 One could also compute the effects of various modulations of the input, Poisson or otherwise, on the ISI distribution.

This is a good place to mention that although the diffusion coefficient in expressions, e.g., (2.4) may depend on t through the process, when the noise is 'multiplicative' the noise variance is often, nevertheless, taken as a constant.

$\mathcal{PP}$2.2.3 There has been little study of the impact of this simplification.

The literature in mathematical neuroscience, in particular about LIF neuron models, has largely been written by researchers with a physics background. Hence we see the terms 'Fokker-Planck' or 'master' equation, usually meaning Kolmogorov equation. The Kolmogorov equation for the LIF model (O-U), between firings, is

$$\frac{\partial P(V, t)}{\partial t} = \frac{\partial}{\partial V}(A(V)P(V, t)) + \frac{1}{2}\frac{\partial^2}{\partial V^2}(B(V)P(V, t)), \qquad (2.5)$$

where the drift coefficient function $A(V) = (1/\tau)(V - V_0 - \mu)$ and the diffusion coefficient $B(V) = 2\sigma^2/\tau$.

The LIF model is said to 'fire' when the O-U-type process (2.4) crosses some constant boundary. This definition of 'firing' is too rigid, however, as compared to other popular neuron firing models, which we will shortly consider. In fact, a 'soft' or stochastic firing rule can be constructed so that firing at a threshold occurs with a probability depending in some way upon the evolution of the process, and possibly also depending on an additional variable. We will see an example of such a rule in Section 2.5 when we look at how an LIF model may be embedded in a 2-variable stochastic Morris-Lecar model as a representative of its behavior during the periods between firings.

A second way in which the usual LIF model may be too rigid is that its voltage is generally reset to a fixed starting value after firing. Other processes go on in a neuron in addition to fluctuation of its voltage. Although the voltage may return to a roughly fixed value after firing, these other aspects may be more appropriately reset to a random rather than to a fixed state, or not reset at all.

$\mathcal{PP}$2.2.4 The problem of how best to do this to mimic the stochastic Morris-Lecar model or 'nature' is open.

2.3   The Izhikevich neuron

An adaptable model was introduced by Izhikevich [20]. This model is like an LIF model in that it does not model spikes as part of a continuous dynamical system as in Sections 2.4, 2.5, or 2.6, but rather resets to a certain voltage after a threshold voltage (30 mV) is reached. Izhikevich's objective was to construct a model that, by parameter choice, can reproduce the spiking behavior of a wide class of neurons, in particular those of 'regular-firing' pyramidal neurons and 'fast-firing' interneurons (usually with inhibitory synapses onto pyramidal neurons). It mimics the behavior of some other models we will discuss next, including their ISI distributions, while being more computationally efficient than they are. The basic deterministic equations are

$$\frac{dV}{dt} = 0.04V^2 + 5V + 140 - u + I \tag{2.6}$$

$$\frac{du}{dt} = -a(bV - u) \tag{2.7}$$

with resetting when $V \geq 30$ *mV*: V is reset to c and u is reset to $u + d$.

Here V represents membrane voltage and u is a recovery variable accounting for activation of the various restoring ionic currents. I represents current input from synapses or other sources. The parameter a is the time scale of the recovery variable u, b is the sensitivity of u to subthreshold fluctuations of V, c is the value to which V is reset after a spike. By choosing parameter values in the appropriate ranges indicated in Table 2.1, one can create models that generate the spiking patterns of regular pyramidal neurons, fast-spiking interneurons, chattering neurons, bursting neurons, thalamocortical neurons, resonator neurons, and so forth [20].

Table 2.1 Parameter values used in (8) and (9) to simulate the Izhikevich model for various neuron types. RS = Regular Spiking; FS = Fast Spiking (inhibitory); IB = Intrinsically Bursting; CH = CHattering; TC = Thalamo-Cortical; RZ = Resonator; LTS = Low Threshold Spiking (inhibitory).

Model	a	b	c	d
RS	0.02	0.20	−65	8.00
FS	0.10	0.20	−65	2.00
IB	0.02	0.20	−55	4.00
CH	0.02	0.20	−50	2.00
TC	0.02	0.25	−65	0.05
RZ	0.10	0.25	−65	2.00
LTS	0.02	0.25	−65	2.00

In practice the Izhikevich neuron model is not used in the deterministic mode - i.e., with I in (2.6) a constant or a simple function. Instead the model typically is made stochastic with modeled noisy synaptic input from e.g., thalamus (among other roles, a sensory 'relay' in the upper brainstem). To do this I is made stochastic, e.g., a Poisson-type input as in LIF, which can be approximated by increments of Brownian motion. Or increments of Brownian motion can be included in (2.7), motivated by the ion channel fluctuations that are part of the 'recovery' process.

$\mathcal{PP}$2.3.1 The question of the relative effects of channel vs. synaptic noise on the output of such a neuron has not been explored.

$\mathcal{PP}$2.3.2 The behavior of a stochastic Izhikevich model will be related to that of the type of stochastic model it should mimic according to the choice of parameters. This has not been explored to our knowledge, although noisy input is routinely used in simulations [20, 21]. A network of Izhikevich neurons is considered in Section 4.5.

2.4   The Hodgkin-Huxley neuron

The primary Hodgkin-Huxley (HH) equation for the firing of a neuron [22], says simply that the rate of change of the membrane potential, V, is proportional to the input, I, minus the output of current through the passing of sodium (Na) and potassium (K) ions through the cell membrane, and leak (L) as in the LIF model:

$$C\frac{dV}{dt} = I - \left[g_{Na}(V - V_{Na}) + g_K(V - V_K) + g_L(V - V_L)\right] \tag{2.8}$$

where C is membrane conductance, I is applied current, g_{Na}, g_K, and g_L are sodium, potassium, and leak conductances, and V_{Na}, V_K, and V_L are 'reversal potentials' that

Table 2.2 Parameter values
used to simulate the HH
model.

Variable	Value	Units
C	1	$\mu F/cm^2$
g_L	0.3	mS/cm^2
$\bar{g}_{Na}$	120	mS/cm^2
$\bar{g}_K$	36	mS/cm^2
V_L	10.6	mV
V_{Na}	115	mV
V_K	-12	mV
ρ_K	18	$1/\mu m^2$
ρ_{Na}	60	$1/\mu m^2$
N_K	$\rho_K \times area$	dimensionless
N_{Na}	$\rho_K \times area$	dimensionless

play the role of minima. There is no threshold as in the LIF model. The usual parameter values are given in Table 2.2. Three additional equations describe the opening and closing of ion channels.

The sodium and potassium conductances are written

$$g_{Na} = m^3 h \bar{g}_{Na}$$
$$g_K = n^4 \bar{g}_K \tag{2.9}$$

where $\bar{g}_{Na}$ and $\bar{g}_K$ are the maximum sodium and potassium conductances. The $m, h,$ and n are called voltage gating variables, and satisfy equations

$$\frac{dx}{dt} = \alpha_x(V)(1-x) - \beta_x(V)x, \quad x = m, h, n. \tag{2.10}$$

Each x can be thought of as a proportion of ion channels of a certain type that are open at time t. Thus, each equation asserts that the rate of change of this proportion is a rate of opening times the proportion of closed channels, minus the rate of closing times the proportion of open channels. Notice that these rates are explicit functions of the membrane voltage, V, and thus the channel openings and closings are 'voltage gated.' At this point we feel optimistic about getting an intuitive understanding of the dynamics of the HH system. This optimistic feeling may be diminished somewhat when we contemplate the set of rate functions α_x, β_x that go into the set of equations (2.10), namely

$$\alpha_m(V) = \frac{0.1(V+40)}{1 - e^{-(V+40)/10}}, \quad \beta_m(V) = 4e^{-(V+65)/18},$$

$$\alpha_h(V) = 0.07e^{-(V+65)/20}, \quad \beta_h(V) = \frac{1}{1 + e^{-(V+35)/10}}, \tag{2.11}$$

$$\alpha_n(V) = \frac{0.01(V+55)}{1 - e^{-(V+55)/10}}, \quad \beta_n(V) = 0.125e^{-(V+65)/80}.$$

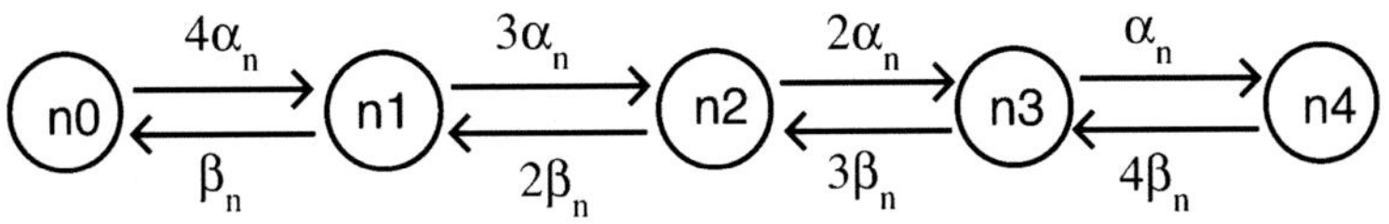

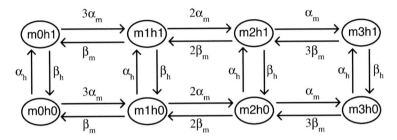

Fig. 2.3 Gate states in a typical potassium channel of the HH neuron model. n*i* represents the number of potassium gates open at a given time in the channel. Arrows are labeled with the transition rates between the n*i* states. Equation 2.9 from [25]

Fig. 2.4 Gate states in the sodium channel of the HH neuron model. m*i*h*j* represents the numbers of sodium activation gates (m*i*) and inactivation gates (h*j*) open in the channel at a given time. Equation 2.10 from [25]

We include this detail in order to give the reader an impression of the level of complexity of the HH model, which won the authors a Nobel Prize in 1963.

To obtain a stochastic HH model one might simply replace the rates α_x, β_x in (2.11) by stochastic rates, by which we mean that the functions α_x, β_x should be used as the rates of Poisson processes where the probability of the event that an ion channel of type x opens in a short time interval dt is $\alpha_x(V_t)\ dt$. In the algorithm [6, 23, 24] that is used, each potassium channel has four gates, all of which must be open for the channel to be open. The scheme is shown in Figure 2.3. A sodium channel has three activation ($m-$)gates and an inactivation ($h-$)gate. It conducts current only when all activation gates are open, and the inactivation gate is closed, i.e., in the $m3h0$ state of the scheme shown in Figure 2.4.

To implement a simulation based on this idea, we write the sodium and potassium conductances as

$$g_{Na} = \frac{Q_{Na}}{N_{Na}}\bar{g}_{Na}, \quad g_K = \frac{Q_K}{N_K}\bar{g}_K, \tag{2.12}$$

where Q_{Na}, Q_K are the stochastic processes expressing the numbers of open sodium and potassium channels at time t, N_{Na}, N_K are the total numbers of sodium and potassium channels involved,

$$N_{Na} = \rho_{Na} \times area, \quad N_K = \rho_K \times area, \tag{2.13}$$

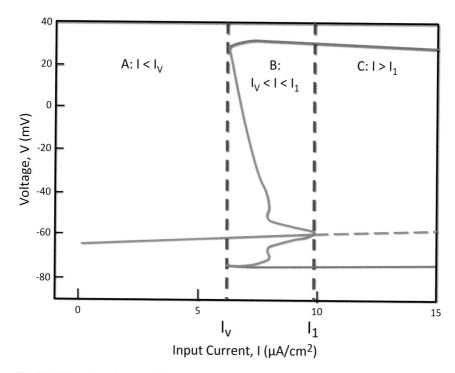

Fig. 2.5 Bifurcation diagram of the Hodgkin-Huxley model neuron. For $I < I_V$ there is a stable fixed point. For $I_V < I < I_1$ there is a stable fixed point surrounded by an unstable limit cycle (orange), which is surrounded by a stable limit cycle (green). For $I > I_1$ the limit point is unstable (blue-dashed) and the stable limit cycle remains. Based upon [25]

and ρ_{Na}, ρ_K are the sodium and potassium channel densities, and *area* is the membrane area. This simulation has been carried out by Rowat [25] to produce a great deal of information about the HH model.

To get an idea of how the HH model produces firing of a neuron, we look at the bifurcation diagram of the deterministic model (Figure 2.5), where the applied current, I, is the bifurcation parameter. The model has a fixed point that is nearly constant in I and is stable for $I < I_1$. For $I > I_1$ the fixed point is unstable. For $I_V < I < I_1$ there is an unstable limit cycle indicated by the orange curve. For $I > I_V$ there is a stable limit cycle outside the unstable one. The model is said to be bistable in the interval $I_V < I < I_1$. The deterministic model started in the domain of attraction of the stable limit cycle spirals toward that cycle.

The circuit of the stable limit cycle is called 'firing.' After one circuit, firing will continue regularly. If the process is started 'inside' the unstable limit cycle it spirals toward the fixed point. The stochastic model combines these two behaviors and alternately circles the stable limit cycle, e.g., firing, a random number of times, and then is quiescent, roughly circling the stable fixed point for a random time before firing again as illustrated in Figure 2.7B for a 2-dimensional Morris-Lecar model.

The stochastic firing pattern of the Morris-Lecar model has been studied in [26], and some of these results are described in the next section.

$\mathcal{PP}$2.4.1 Such a study, however, has not been done for the stochastic HH. Several of the results might well be similar. Because the HH model is 4-dimensional, however, one cannot simply speak of 'inside' and 'outside' the unstable limit cycle, and this will pose new problems in describing the results.

Much effort has gone into devising stochastic differential equation schemes for simulating stochastic HH, because such schemes are much faster than simulation of the discrete state space Markov chain as described above, by what has come to be called the Gillespie algorithm [23]. Some of these schemes are compared in Figure 1.2. In this figure, Micro uses the Gillespie algorithm, Orio is the algorithm that comes from the diffusion approximation of the precise piecewise Poisson process by the method of Kurtz [18], Guler, Linaro, and Fox are other SDE-based schemes, of which the Guler scheme matches the micro-scheme the best. The stochastic schemes for HH make possible a detailed simulation study of the ISI distribution as it depends on both deterministic input, I, and stochastic input. The stochastic input, coming from ion channels, is proportional to $A^{-1/2}$, where A is the membrane area, in (2.13). In Figure 1.2 we see that as area increases, and variance decreases, sub-maxima in the ISI distribution become prominent. The initial peak corresponds to the time taken by neuron firing itself.

$\mathcal{PP}$2.4.2 The resulting information about the ISI distribution in [8] makes a further study of the sample-path behavior of HH very inviting.

In fact, HH has been neglected as an object of study until recently, because of its complexity. Now it has become reachable with computation methods, which may suggest analytical conjectures. We will be able to discuss HH further after we introduce the stochastic Morris-Lecar model. In particular we will suggest explanations for some of the features of the ISI distribution as shown in Figure 1.2.

2.5 The Morris-Lecar neuron

The Morris-Lecar (ML) model neuron is a simplification of the Hodgkin-Huxley model. There are only two equations, the primary one, very similar to the voltage equation (2.8) of the HH model, telling how the voltage changes,

$$C\frac{dV}{dt} = I - \bar{g}_{Ca}m_\infty(v)(v - V_{Ca}) - \bar{g}_K w(v - V_K) - g_L(v - V_L), \qquad (2.14)$$

and an equation for the fraction of open potassium channels, similar to the x-equation (2.10) for HH,

$$\frac{dw}{dt} = \alpha(v)(1 - w) - \beta(v)w, \qquad (2.15)$$

Table 2.3 Parameter values
used to simulate the ML
model.

Variable	Value	Units
C	20	$\mu F/cm^2$
g_L	2.0	mS/cm^2
$\bar{g}_{Ca}$	4.4	mS/cm^2
$\bar{g}_K$	8	mS/cm^2
V_L	−60	mV/cm^2
V_{Ca}	120	mV/cm^2
V_K	−84	mV/cm^2
V_1	−1.2	mV/cm^2
V_2	18.0	mV/cm^2
V_3	2.0	mV/cm^2
V_4	30.0	mV/cm^2
ϕ	0.04	dimensionless

where

$$m_\infty(v) = 0.5\left(1 + tanh\left(\frac{v - V_1}{V_3}\right)\right),\qquad(2.16)$$

and $\alpha(v)$ and $\beta(v)$ are given by

$$\alpha(v) = 0.5\,\phi\,cosh\left(\frac{v - V_3}{2V_4}\right)\left(1 + tanh\left(\frac{v - V_3}{V_4}\right)\right),\qquad(2.17)$$

$$\beta(v) = 0.5\,\phi\,cosh\left(\frac{v - V_3}{2V_4}\right)\left(1 - tanh\left(\frac{v - V_3}{V_4}\right)\right).\qquad(2.18)$$

The ML model was constructed to describe the motor neuron action of the barnacle muscle [27], and has been widely adopted by neuroscientists as a useful model of many real neurons though only two-dimensional. We focus here on the so-called Type II ML neuron, where the parameter range yields a bifurcation diagram (Figure 2.6) similar to that of HH (Figure 2.5). The parameter set used here is given in Table 2.3.

The stochastic ML model is obtained by interpreting the rate functions, $\alpha(v), \beta(v)$ in (2.15) as Poisson rates. Then the process (V_t, w_t) can be simulated using the Gillespie algorithm, which simulates a sample path by creating waiting times alternately with jumps using proportional jump probabilities. If we prefer to use an SDE simulation, we can write an SDE approximation using the result of Kurtz [18]. The resulting SDE system will consist of the same voltage equation, (2.14), together with

$$dw = \big(\alpha(v)(1 - w) - \beta(v)w\big)dt + \frac{1}{\sqrt{N_K}}\big(\alpha(v)(1 - w) + \beta(v)w\big)^{1/2}dW\qquad(2.19)$$

where N_K is the number of potassium channels in our system.

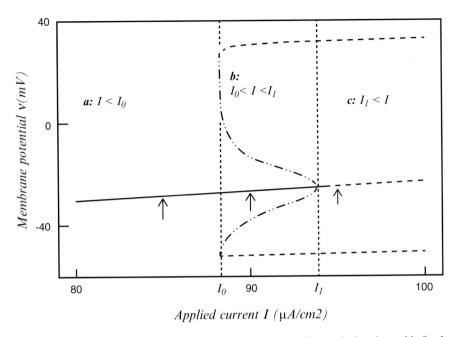

Applied current I (μA/cm2)

Fig. 2.6 Bifurcation diagram of the Morris-Lecar model neuron. For $I < I_0$ there is a stable fixed point. For $I_0 < I < I_1$ there is a stable fixed point surrounded by an unstable limit cycle, which is surrounded by a stable limit cycle. For $I_1 < I$ the limit point is unstable and the stable limit cycle remains. Compare with Figure 2.5. Reprinted with permission from [26]

The dynamics of the deterministic and stochastic ML models are illustrated in Figure 2.7 for $I = 82, 90,$ and $98 \, \mu A/cm^2$, where the simulation uses the Gillespie algorithm.

Figure 2.7 should be compared with Figure 2.8 where the simulation uses the SDE and the Euler method. It is apparent that the two very different methods of simulation create substantially the same picture.

The stochastic ML fires by traversing the stable limit cycle some number of times, alternating with spending some quiescent time near the stable fixed point. Questions that are partially answered in [26] are: what is the distribution of the number of consecutive firings, and what is the distribution of the length of quiescent times between firings, i.e. the ISI distribution? The number of consecutive firings before a quiescent period has a geometric distribution with parameter $p = p(I)$, the probability that a circuit of the limit cycle is followed by a quiescent period. This is made clear by simulations together with an argument that successive firings correspond to independent segments of path.

$\mathcal{PP}$2.5.1 But there is no proof of this.

The tail of the ISI distribution is exponential by the argument that the time at which a Markov process, which is in equilibrium in a region, escapes from the region is exponential. During the first few circuits around the fixed point after a firing the

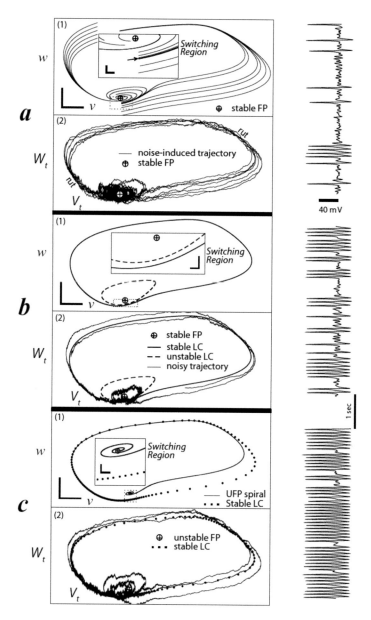

Fig. 2.7 Dynamics of the Morris-Lecar model neuron. Parts (1),(2) of each subfigure are the deterministic, stochastic model. In (a), (b), (c) the input voltage, I, is 82, 90, 98 mV/cm^2. The large loops of the path correspond to the firing of the neuron model. The small excursions around the fixed point correspond to subthreshold oscillations in quiescent periods. The side strip shows sample path projections onto the V-axis. As I increases more time is devoted to firing and less to subthreshold oscillations. Note that deterministic dynamics differs in regions of I represented by (a), (b), (c), but stochastic dynamics is rather similar, an example of smoothing by noise. Reprinted with permission from [26]

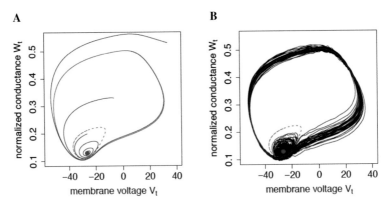

Fig. 2.8 Dynamics of the (A) deterministic ML model, and (B) stochastic model simulated from the SDE using the Euler method. Compare with Figure 2.7, where the Gillespie algorithm is used to simulate the discrete state space Markov chain version of the ML model neuron. The two computational methods appear to illustrate the dynamics equally well. Reprinted with permission from [28]

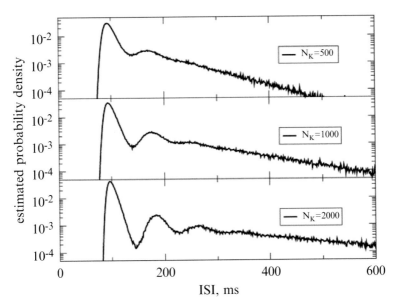

Fig. 2.9 ISI histograms of the Morris-Lecar model neuron. Reprinted with permission from [26]

stochastic ML process is not in an equilibrium inside the unstable limit cycle, but is more likely to escape from it and fire again at each pass near the orbit of the stable limit cycle (see Figure 2.7). This roughly periodic pattern is reflected in the oscillations seen in the ISI histogram of the stochastic ML neuron (Figure 2.9), after the initial peak that estimates the distribution of the time taken by a single firing. A similar pattern is seen in the ISI histogram of the stochastic HH neuron (Figure 1.2).

$\mathcal{PP}$2.5.2 The methods used in [26] to obtain these results combine simulation and stochastic analysis. Their conversion to mathematical results will require additional stochastic ideas and analysis.

The ISI histogram for ML (Figure 2.9) indicates that ML has several features in common with HH. First, there is an initial peak associated with runs of firings with no intervening quiescent period. The second peak is generated by ISIs where exactly one subthreshold oscillation occurs between two firings. The third peak, almost imperceptible in this figure, is generated by ISIs with exactly two subthreshold oscillations. The tail of the ISI distribution is exponential, by a familiar argument, and the parameter of that exponential has been estimated using simulations.

$\mathcal{PP}$2.5.3 But can someone compute that exponential parameter?

$\mathcal{PP}$2.5.4 Can someone prove the explanations just given for the local maxima in the ISI plot for ML, and similar explanations in the case of HH, which seem compelling but are really just strong conjectures?

An insight about the relation of the LIF model to ML is contained in [28]. There we learn what the title says, that the quiescent period between firings of an ML model is stochastically (almost) an O-U process. The way to see this is to linearize the deterministic ML dynamics about the fixed point to obtain a 2-dimensional linear system

$$\frac{dY}{dt} = \mathbb{A}Y. \tag{2.20}$$

The matrix $\mathbb{A}$ has complex eigenvalues $-\lambda \pm i\omega$ with negative real part and $0 < \lambda \ll \omega$, and hence the deterministic model has damped oscillations. The diffusion process,

$$dZ = \mathbb{A}Z dt + \mathbb{N}dW_t, \tag{2.21}$$

in which the oscillations are sustained by the stochastics, is called a process of quasicycles [29]. In [30] it is shown that a diffusion of the form (2.21) is well approximated, when $0 < \lambda \ll \omega$, by the diffusion

$$\tilde{Z} = \frac{\sigma}{\sqrt{\lambda}} \mathbb{Q} R_{-\omega t} S_{\lambda t}, \tag{2.22}$$

where $\mathbb{Q}$ is a matrix that transforms $\mathbb{A}$ in Eqn. (2.21) into a canonical form

$$\mathbb{Q}^{-1}(-\mathbb{A})\mathbb{Q} = \begin{pmatrix} -\lambda & \omega_d \\ -\omega_d & -\lambda \end{pmatrix} := \mathbb{A}_1, \tag{2.23}$$

$\mathbb{R}$ is the rotation

$$\mathbb{R}_s = \begin{pmatrix} \cos(s) & -\sin(s) \\ \sin(s) & \cos(s) \end{pmatrix}, \tag{2.24}$$

$\mathbb{S}(t)$ is a standard two-dimensional Ornstein-Uhlenbeck process with independent components, and

$$\sigma = \sqrt{0.5\mathrm{Tr}(\mathbb{Q}^{-1}\mathbb{N}\mathbb{N}^{\mathsf{T}}(\mathbb{Q}^{-1})^{\mathsf{T}})} \tag{2.25}$$

is a scalar.

The matrix $\mathbb{N}$ can be evaluated at the stable fixed point of the deterministic system (2.14), (2.15), because this computation is for ML dynamics in the neighborhood of this fixed point. The approximation of (2.21) by (2.22) is proved using the Martingale problem method, following several changes of variables. We sketch this proof here, both for clarity and also because the approximation is important in Chapter 3 as well.

We make three changes of variables in (2.21). First we transform the matrix $\mathbb{A}$ to normal form in order to see more clearly the separate effects of the relatively slow damping characterized by λ, and the relatively fast rotation characterized by ω. Using a matrix $\mathbb{Q}$ as in (2.23), we write $\mathbb{Y}(t) = \mathbb{Q}^{-1}\mathbb{V}(t)$. Then

$$d\mathbb{Y}(t) = \mathbb{A}\mathbb{Y}(t) - \mathbb{C}d\mathbb{W}(t), \tag{2.26}$$

where $\mathbb{C} = \mathbb{Q}^{-1}\mathbb{N}$. The noise in (2.26) has covariance matrix

$$\mathbb{B} := \mathbb{C}\mathbb{C}^{\mathsf{T}} = \mathbb{Q}^{-1}\mathbb{N}\mathbb{N}^{\mathsf{T}}(\mathbb{Q}^{-1})^{\mathsf{T}}. \tag{2.27}$$

Next we write

$$\mathbb{Y}(t) = \mathbb{R}_{\omega t}\mathbb{Z}(t), \tag{2.28}$$

where $\mathbb{R}_s$ is the rotation (2.24). Using the SDE (2.26) and Itô's formula we obtain

$$d\mathbb{Z}(t) = -\lambda \mathbb{Z}(t)dt + \mathbb{R}_{\omega t}\mathbb{C}d\mathbb{W}(t). \tag{2.29}$$

Finally, in order to compare the process $\mathbb{Z}(t)$ with a standard two-dimensional Ornstein-Uhlenbeck process, we rescale time and space by writing

$$\mathbb{U}(t) = \frac{\sqrt{\lambda}}{\sigma}\mathbb{Z}(t/\lambda), \tag{2.30}$$

where

$$\sigma^2 = \frac{1}{2}\mathrm{tr}(\mathbb{B}) = \frac{1}{2}(B_{11} + B_{22}) = \frac{1}{2}\sum_{i,j=1}^{2} C_{i,j}^2. \tag{2.31}$$

To identify the SDE that $\mathbb{U}(t)$ satisfies it is convenient to use the integrated form

$$\mathbb{U}(t) - \mathbb{U}(0) = \frac{\sqrt{\lambda}}{\sigma}\left(-\lambda \int_0^{t/\lambda} \mathbb{Z}(s)ds + \int_0^{t/\lambda} \mathbb{R}_{\omega s} \mathbb{C} d\mathbb{W}(s) \right)$$

$$= \frac{\sqrt{\lambda}}{\sigma}\left(-\int_0^t \mathbb{Z}(u/\lambda)du + \frac{1}{\sqrt{\lambda}} \int_0^t \mathbb{R}_{\omega u/\lambda} \mathbb{C} d\tilde{\mathbb{W}}(u) \right), \qquad (2.32)$$

where $\tilde{\mathbb{W}}(t) = \sqrt{\lambda}\, \mathbb{W}(t/\lambda)$ is a standard Brownian motion. Therefore

$$d\mathbb{U}(t) = -\mathbb{U}(t)dt + \mathbb{R}_{\omega t/\lambda} \mathbb{D} d\tilde{\mathbb{W}}(t), \qquad (2.33)$$

where $\mathbb{D} = 1(1/\sigma)\mathbb{C}$, so that $\mathrm{tr}(\mathbb{D}\mathbb{D}^\top) = 2$.

Theorem 1 of [30] says that the distribution of $\{\mathbb{U}^\lambda(t) : 0 \le t \le T\}$, where $\mathbb{U}^\lambda$ denotes the process satisfying (2.33) with a fixed λ, ω, $\mathbb{U}^\lambda(0) = x$, fixed, and $0 < T$ fixed, converges, as λ/ω goes to 0, to the distribution of the standard 2-dimensional O-U process $\{\mathbb{S}(t) : 0 \le t \le T\}$ with $\mathbb{S}(0) = x$.

The approximation (2.22) is obtained by reversing the three changes of variables starting from the SDE for the standard 2-dimensional O-U process

$$d\mathbb{S}(t) = -\mathbb{S}(t)dt + d\mathbb{W}(t). \qquad (2.34)$$

$\mathcal{PP}$2.5.5 In order to obtain an error bound on the approximation one would need a uniformity result with $T \to \infty$ as $\lambda/\omega \to 0$, in Theorem 1 of [30].

In an application to the Wilson-Cowan type system (see Chapter 3), we have shown in [3] that the approximation (2.22) is quite good for the parameters we consider (see also [30]).

The amplitude (modulus) of the process $\tilde{Z}$ satisfying equation (2.22) is proportional to the amplitude process of the standard 2-dimensional O-U process,

$$A_{\lambda t} = \sqrt{(S_{\lambda t}^1)^2 + (S_{\lambda t}^2)^2}. \qquad (2.35)$$

This process solves the stochastic differential equation

$$dA_t = \left(\frac{1}{2A_t} - A_t \right)dt + dW_t, \qquad (2.36)$$

see, e.g., [31]. The fact that the amplitude of the ML stochastic model, (2.36), is (almost) an O-U process is the basis of the claim that the ML model embeds an LIF model in its quiescent periods between firings [28]. It is evident from simulations, e.g., Figure 2.7, that the pieces of sample path during successive firings are very nearly identical. Hence the stochasticity in the system is not affecting the dynamics very much during the times of firing, which are, in addition, rather fast relative to the circuits of the fixed point during quiescence. Two additional questions arise in the comparison of LIF with ML. How do firing and reset occur? In terms of the ML how does the switching from quiescence to firing and from firing to quiescence occur? In the LIF model there is a crisp threshold crossing and reset, often to zero, the center point of the O-U. If we believe that the ML captures the essential dynamics of the system, then both firing and reset are wrongly defined in LIF. In [28] an alternative definition of firing in terms of a hazard and rate function is proposed, and its adoption seems to move the LIF in the direction of the ML in terms of the ISI distribution.

$\mathcal{PP}$2.5.6 A more clever stochastic analysis, however, might produce a better or even the 'correct' solution.

$\mathcal{PP}$2.5.7 The question of a suitable reset rule is similarly open. A natural thought is that the reset should not be to 0, or to a fixed point, but rather should be to a value inspired by the observation that the ML enters quiescence by crossing to the inside of the unstable limit cycle.

An important aspect of real neurons that is captured by ML and HH neuron models is subthreshold oscillations. Neither an LIF model nor an Izhikevich model generates such oscillations.

$\mathcal{PP}$2.5.8 A stochastic model could be constructed, similar to the Izhikevich model in having a firing threshold and two equations, and hence fast to simulate, and at the same time incorporating subthreshold oscillations sustained by noise, i.e., quasicycles.

Having had a look at the stochastic dynamics of ML, we are now in a position to think about the relation of ML to HH. In general, open problems about ML are even more open for HH. On the other hand, our knowledge about ML from, e.g., [26, 28], combined with knowledge about HH such as details about its ISI distribution from [8], leads us to further insight into the forbiddingly complex stochastic HH model.

If we compute the fixed point for deterministic HH, and linearize this 4-dimensional system about the fixed point, obtaining a linear system $dY/dt = \mathbb{A}Y$, the matrix $\mathbb{A}$ will have a pair of complex eigenvalues $-\lambda \pm i\omega$. The other two eigenvalues will be negative and larger than λ in magnitude. We 'know' this without doing the computation because the ISI distribution of HH (Figure 1.2) has local maxima after the initial peak, just as does the ISI distribution of ML (Figure 2.9). This indicates that after a 'last' firing the stochastic HH path returns to quiescence at a point just 'inside' the unstable limit cycle, which we see in the bifurcation diagram, so that it is quite likely to fire again after one cycle around the stable fixed point, hence the first local maximum after the initial ISI peak. Why should the language of the 2-dimensional model be appropriate for the 4-dimensional model? Because the remaining two eigenvalues will tell us that the stable ISI process will rapidly damp from wherever it is to the hyperplane defined by the eigenvectors corresponding to the complex eigenvalues. When we project the bifurcation diagram to this hyperplane, the language of 'just inside the stable limit cycle' does make sense.

$\mathcal{PP}$2.5.9 So the picture is there, and it remains to make the story into mathematics.

When we have a quantitative hold on the model, whether HH or ML, we will be able to use it to learn more about the dynamics of real neurons.

A striking property of stochastic ML that no doubt holds also for stochastic HH is that the stochastic model does not 'see' the endpoints of the bistability region. In other words, parameters of the stochastic dynamics controlling the lengths of the firing and quiescent cycles are continuous across these boundaries, which are discontinuity points of the deterministic dynamics. Figure 2.10 shows switching probabilities of stochastic ML, p being the probability that firing stops and r being the probability that firing starts (see caption of Figure 2.10) for a range of input current levels, I, extending far on both sides of the bistability range, and for various values of N_K, the number of potassium channels, which decreases with increasing variance. What is striking is the smoothness of these curves across both endpoints of the bistability region. This indicates that the states of firing and quiescence, corresponding in the deterministic model to the stable limit cycle and the stable limit point in the bistability region of input, I, continue to play exactly the same roles in the stochastic dynamics over a large additional range of I. This has been shown for ML by a simulation study in [26].

$\mathcal{PP}$2.5.10 The corresponding stochastic analysis is missing.

$\mathcal{PP}$2.5.11 Both simulation and stochastic analysis are missing in the HH case, although the result almost certainly holds in this case also.

2.6 The FitzHugh-Nagumo neuron

Another simplification of the HH model, now called the FitzHugh-Nagumo (FH-N) model, was proposed by FitzHugh [32] and later modified by Nagumo and colleagues [33]. While trying to analyze the phase-space dynamics of the HH model, FitzHugh noticed that neural action potentials are very similar to heart beats, which had been described as relaxation oscillations in a model by van der Pol and van der Mark [34]. The model FitzHugh proposed for neurons was

$$\frac{dV}{dt} = \alpha(w + 3V - V^3 + I) \tag{2.37}$$

$$\frac{dw}{dt} = -(V - a + bw)/\alpha, \tag{2.38}$$

where V, again, is voltage and w, like w in (2.15) of Morris-Lecar, stands for the gating variables of Hodgkin-Huxley, and w is called the 'recovery variable.' Equation (2.37) models a fast process, basically the action potential, and equation (2.38) models all of the ion channel activity as a single recovery variable w embedded in a slow linear process parametrized by a and b. The input, I, drives the voltage V which cycles through buildup and release. The cubic term $-V^3$ bounds the buildup and forces the relaxation that is parametrized by α.

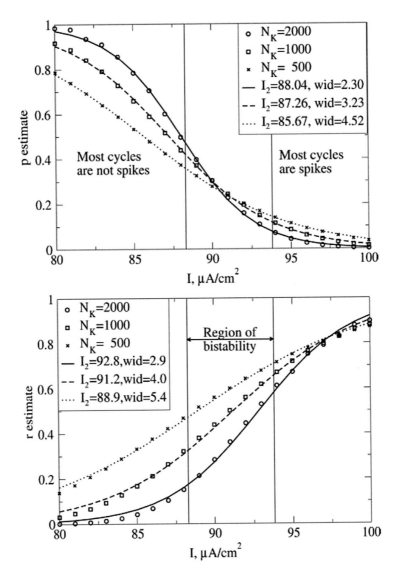

Fig. 2.10 Continuity of parameters controlling the pattern of firing and quiescence across the bistability region of the stochastic ML model neuron. $p = p(I)$ is the probability that a traversal of the limit cycle is followed by switching to the neighborhood of a fixed point. $r = r(I)$ is the probability that a tight circuit of the fixed point (inside the unstable limit cycle when it exists) is followed by a traversal of the limit cycle or a large circuit. Examination of Figure 2.7 reveals that $p(I)$ and $r(I)$ are well-enough defined to allow computation of statistics from simulations. Reprinted with permission from [26]

The FH-N model has been used in many biophysical applications because it faithfully reproduces the spiking behavior of the HH model, although it does not deal with the ionic currents and so cannot directly model channel noise. One of the most striking applications of the stochastic FH-N model has been in demonstrating stochastic facilitation (stochastic resonance) in model neurons, first done by Longtin [14]. Longtin modified the FH-N equations by adding stochastic forcing to the fast equation and a slow periodic forcing to the recovery equation (see also Section 1.2). He justified this by the observation that the time scale of synaptic noise is of the order of the action potential itself, whereas a periodic signal to the neuron, such as a local field potential oscillation at say 10 Hz, would be of the order of the recovery variable time scale. Alternatively, we could add noise to the recovery equation to model ion channel noise alone (in spite of the much faster time scale of the ion channel noise).

Yet another way to create a stochastic FH-N model is to add noise to both equations, perhaps to include synaptic and ion channel noise in the same system but to separate their effects. This is done by Berglund and Landon in [35]. They construct a definition of 'the number of sub-threshold circuits of the fixed point of the FH-N model between firings,' and then show that the distribution of the number is asymptotically geometric.

$\mathcal{PP}$2.6.1 This could be, but hasn't been, done for the ML model.

On the other hand, the number of consecutive firings of ML has been shown by simulation in [26] to be geometrically distributed.

$\mathcal{PP}$2.6.2 A proof of this is lacking for both FH-N and ML.

$\mathcal{PP}$2.6.3 Another problem is to show in what sense the FH-N model approximates the ML model.

Perhaps the two stochastic models are more similar than the two deterministic ones are. Figure 2.11 shows the sample path behavior of a stochastic FH-N model. The oscillation on the left is analogous to the oscillation near the fixed point of ML in Figure 2.8B.

2.7   What can a single neuron compute?

As stated in the introduction, a 'formal' neuron has an input, internal dynamics, and an output. A single neuron can be said to compute a transfer function, turning input into output. The exact nature of this transfer function can be quite important in neural networks containing only a few neurons, for example in the parasitic worm *C. elegans*, whose nervous system contains only a few hundred neurons. In such networks each neuron performs a specific role in controlling behavior such as mating, feeding, and learning.

First, and most naturally, the output can be a more or less faithful copy of the input, especially as regards the more prominent aspects of that input, such as a sinusoidal oscillation. This 'relay' computation is the one most often assumed by the early modelers of stochastic facilitation mentioned in Section 1.2. Second, a neuron

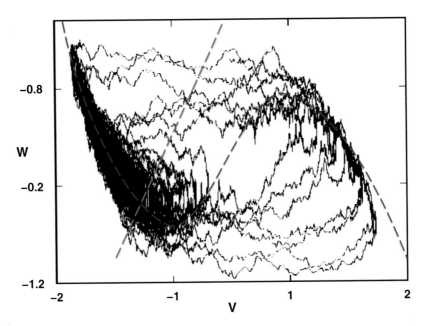

Fig. 2.11 Dynamics of the stochastic FH-N neuron. The colored dashed lines are the nullclines. Large box-like loops correspond to firings and smaller circuits at left correspond to subthreshold oscillations. The separation between forcing and subthreshold oscillations here is not so clear as that in the stochastic ML neuron (Figures 2.7 and 2.8). Reprinted with permission from [36] with nullclines added

can act as a filter, or resonator, in that certain input frequency bands are privileged in their output. Such action usually reflects the frequency of subthreshold oscillations. A neuron also integrates (usually adds linearly up to saturation) all of its inputs over space (cell surface) and time (usually over a period of 10 ms to 100 ms), particularly in its dendrites. Finally, a neuron develops a synaptic weighting distribution on its inputs as a function of its synaptic transmission history, using mechanisms such as spike-timing-dependent plasticity (synaptic changes dependent on timing of input from other neurons) and long-term potentiation (synaptic changes caused by repeated inputs from other neurons). Thus its transfer function varies over time under the influence of its history and its neural environment. Other mechanisms affecting the transfer function include gene expression effects on neurotransmitter receptors, effects of glia (cells that nourish and otherwise support neuron function), and effects of the river of ion-rich extracellular fluid in which each neuron lives (e.g., [37, 38]).

Each of the types of transfer function just mentioned can be implemented when needed in simple neural networks.

$\mathcal{PP}$2.7.1 But as far as we know few studies have been made of how to implement the various types of transfer function for the different single neurons models just described, or indeed whether particular models can only implement a subset of them.

$\mathcal{PP}$2.7.2 And also, few studies have been done of which type of single neuron model fits best into the various roles required. One specific model to address would be that of a learning circuit in *C. elegans*, in which only a few neurons each have a very specific function [39].

We discuss this topic further in Section 4.5 in the context of to what extent motifs (specific arrangements of small groups of neurons) matter in population models as a function of their size.

Chapter 3
Population and Subpopulation Models

We have seen stochastic neuron firing models that have inherent frequencies in their subthreshold dynamics, e.g., for the Morris-Lecar neuron [28]. This frequency shows up at the population level. If we record the firings of several model neurons over a period of time, the firings of each single neuron follow an inherent frequency, but often skipping many repeats of that frequency. The skipping phenomenon is a result of the tendency of the neurons to fire on their subthreshold quasicycles, a stochastic facilitation phenomenon explained in Sections 1.2 and 2.5. When the firings of the several neurons, driven by the same or partially common noise, are added together as a function of time, we obtain a function that oscillates at the common inherent frequency of the family.

On the other hand, when we consider a population of interacting, stochastic, neurons the characteristics of their interaction may produce a population oscillatory frequency that could combine with or dominate any inherent frequency of the individual neurons. This section describes stochastic problems arising from populations of interacting neurons as well as from interacting neuron populations.

3.1 Spiking versus rate models

Before we proceed to the characteristics of populations of neurons, we first consider how to move from detailing models of individual neurons, as we did in the previous section, to describing the activities of large populations of neurons such as those found in all mammalian and many other animal brains. The problem was stated clearly by Wilson and Cowan [40]. First, the number of neurons, plausibly in the millions in many cases, is simply too vast for most approaches beginning with the details of individual neurons to be tractable (but see [21]). Second, it is likely to be the global properties of the aggregation of the many neurons that are important for processes such as learning, memory, pattern recognition, and so forth, not the local

© Springer International Publishing Switzerland 2016
P.E. Greenwood, L.M. Ward, *Stochastic Neuron Models*, Mathematical Biosciences
Institute Lecture Series, DOI 10.1007/978-3-319-26911-5_3

properties of individual neurons. Third, even if local interactions are stochastic, they might give rise to very orderly activity when viewed globally. What is important then is not whether any given neuron is firing at a particular time, but the overall frequency, or rate, of firing in a functionally relevant population or subpopulation. For this reason, Wilson and Cowan [40] introduced variables representing the proportions of excitatory and inhibitory binary neurons firing per unit time at each instant t, into their model of the dynamics of a spatially localized population.

Such 'rate models' have been derived from deterministic spiking models of individual neurons by making a set of simplifying assumptions. This has been done in a variety of ways, depending on which type of neuron was taken as a starting point. 'Established' properties of neurons are used to justify these assumptions. For example, every neuron is thought to be subject to a refractory, or delay, period, r, after firing an action potential, so the proportion of binary neurons that fire in the next r ms is a joint function of the proportion of neurons that are not refractory, and of the proportion that receive at least 'threshold' excitation. Further assumptions can be made about how much excitation the various neurons receive, and the distribution of 'thresholds' of the different neurons. Proceeding like this, nonlinear differential equations describing the dynamics of the neural populations have been developed [40]. Similarly, Appendix A of [41] adumbrates the development of the rate model we describe in Section 3.2. They begin with a description of the time evolution of conductance for receptors of excitatory inputs in a neuron, and end up approximating solutions to a pair of integral equations to get expressions for the time derivatives that are the neuron firing rates. Such rate models are quite separate from the population oscillations mentioned in the introduction to this section, which are based on the subthreshold dynamics of single neurons.

Experiments show that neuron population oscillations exist and their power tends to covary with spiking activity during animal behavior. Further, such oscillations are predicted by rate equations as seen in Section 3.2 and [3]. One caveat about this relationship is that experiments also show that the peak power of population oscillations and the corresponding peak spiking activity vary over a wide frequency range, from less than 1 Hz to more than 200 Hz (500 Hz is the theoretical limit of individual neuron firing frequency). Volleys of spikes can occur up to several thousand Hz because each neuron only has to fire occasionally to contribute to a population dynamic (e.g., [4], see Section 3.4). Does this covariation between frequency and peak power in population oscillations hold over this entire range? Apparently not. A recent study found that it holds only for the higher frequencies, above 80 Hz [42]. Thus, so-called low-gamma (30–80 Hz) power may be dissociated from spike rate, whereas high-gamma power always remains strictly bound to spike rate. When working with rate models, as in what follows, we need to keep this finding in mind. Indeed, when the model has a resonant frequency below 80 Hz, as in the model of [3, 41], we are likely modelling the local field potential (LFP, typically the sum of all electrical fields at a given point in the brain excluding spike potentials) of the population but not the population spike rate per se.

3.2 Populations of interacting neurons

Substantially all cortical neurons are either excitatory or inhibitory. Indeed, as pointed out by Wilson and Cowan [40], probably all neural activity of any complexity is dependent on the interaction of excitatory and inhibitory neurons. The words 'excitatory' and 'inhibitory' describe what a neuron does to other neurons. A typical configuration involves one excitatory and one inhibitory neuron in an arrangement illustrated in Figure 3.1, where the excitatory neuron excites itself and the inhibitory neuron, and the inhibitory inhibits itself and the excitatory neuron. Alternatively, E and I in Figure 3.1 can represent populations of neurons of the same types.

A stochastic dynamical system illustrated in Figure 3.1 with sample paths like those in Figure 3.2 is

$$\tau_E dV_E(t) = (-V_E(t) + S_{EE}V_E(t) - S_{EI}V_I(t))dt + \sigma_E dW_E(t)$$

$$\tau_I dV_I(t) = (-V_I(t) - S_{II}V_I(t) + S_{IE}V_E(t))dt + \sigma_I dW_I(t). \tag{3.1}$$

Here V_E, V_I are the voltages or firing rates of excitatory and inhibitory neurons, or families of neurons, τ_E, τ_I are time constants, and σ_E, σ_I are the amplitudes of independent, standard Brownian motions, W_E, W_I. The parameters $S_{EE}, S_{II}, S_{IE}, S_{EI} \geq 0$, are constants that represent the mean efficacies of the excitatory or inhibitory synaptic connections to post-synaptic neurons within each separate population, as indicated by the notation, with S_{IE} representing input to inhibitory from excitatory neurons, and so on. The system (3.1) can be regarded as the local linearization at its fixed point, moved to (0,0), of the Wilson-Cowan model [40] described in Section 3.3.

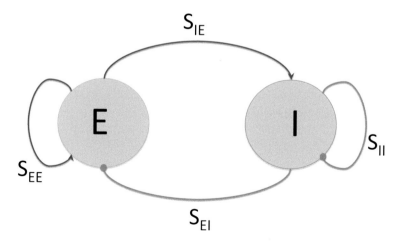

Fig. 3.1 Typical arrangement of an excitatory (E) and inhibitory (I) neuron pair or pair of populations composed of those neuron types. $S_{EE}, S_{II}, S_{EI}, S_{IE}$ refer to the synaptic efficacies of the connections (see (3.1))

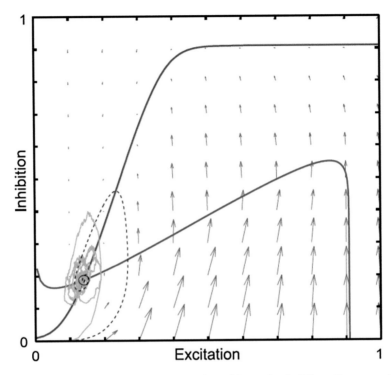

Fig. 3.2 Nullclines and quasicycle phase plane plots of the stochastic Wilson-Cowan population model. The behavior of the Wilson-Cowan model [43] near the fixed point for parameter values where $\lambda < 0$ is similar to that of the ML model near its fixed point, Figures 2.7 and 2.8, and can be described by the approximation (3.6), (3.7), (3.8). Reprinted with permission from [44]

The system (3.1) has, in fact, the same form as the simple linear diffusion (2.21), which describes the local behavior of a Morris-Lecar neuron model near its fixed point. For parameters of interest, the system (3.1) has the same stochastic dynamics as the system (2.21) with oscillations sustained by noise. Next we indicate how to extend the analysis of Section 2.5 to obtain an explanation of a property of LFP data from the visual cortex: there are intermittent bursts of gamma frequency (40–80 Hz) oscillations in response to an uninformative stimulus.

The system (3.1) has already been centered at the fixed point, (0,0), and can be written as

$$dV = -\mathbb{A}Vdt + \mathbb{N}dW \tag{3.2}$$

where $\mathbb{V} = (V_E(t), V_I(t))^\top$, $d\mathbb{W} = (dW_E(t), dW_I(t))^\top$, and

$$\mathbb{A} = \begin{pmatrix} (1 - S_{EE})/\tau_E & S_{EI}/\tau_E \\ -S_{IE}/\tau_I & (1 + S_{II})/\tau_I \end{pmatrix}, \quad \mathbb{N} = \begin{pmatrix} \sigma_E/\tau_E & 0 \\ 0 & \sigma_I/\tau_I \end{pmatrix}. \tag{3.3}$$

As already mentioned, the linear system (3.2) has the same form as (2.21), so that in a suitable parameter range we will have the approximation (2.22).

When we take $\mathbb{N} = 0$, and use the parameters as in [3], the eigenvalues of $-\mathbb{A}$ are $-\lambda \pm i\omega$, with $0 < \lambda \ll \omega_d$. Here, again, λ is the damping rate of the oscillation,

$$\lambda = 0.5 \left[\frac{1 - S_{EE}}{\tau_E} + \frac{1 + S_{II}}{\tau_I} \right], \tag{3.4}$$

and ω_d is its natural frequency,

$$\omega_d = \sqrt{ \frac{S_{EI} S_{IE}}{\tau_E \tau_I} - 0.25 \left[\frac{1 - S_{EE}}{\tau_E} - \frac{1 + S_{II}}{\tau_I} \right]^2 }, \tag{3.5}$$

which is positive when the eigenvalues are complex.

The deterministic system obtained from equation (3.2) with $\mathbb{N} = 0$ has oscillations that damp to the stable point $(0,0)$ at rate λ. The stochastic system in equation (3.2) with $\mathbb{N} > 0$ has sustained oscillations of a narrowband nature [30].

The burstiness of local field potentials (LFPs) can be shown by approximating the linear process (3.1), as in Section 2.5 on the ML neuron, by a process of the form (2.22) by the argument given in (2.20)–(2.34) of Chapter 2. Further details are in [28]. The two-dimensional process $\mathbb{V}(t) = [V_E(t), V_I(t)]$ can be approximated as

$$\mathbb{V}(t) = [V_E(t), V_I(t)]^\top \approx \frac{\sigma}{\sqrt{\lambda}} \mathbb{Q} \mathbb{R}_{-\omega_d t} \mathbb{S}(\lambda t), \tag{3.6}$$

where $|\mathbb{S}(\lambda t)| = \sqrt{S_1(\lambda t)^2 + S_2(\lambda t)^2}$, and $\phi(\lambda t) = \arg[S_1(\lambda t) + i S_2(\lambda t)]$. The process $Z(t) = |\mathbb{S}(t)|$ is called the *radial process* or *modulus process* associated with $\mathbb{S}(t)$, and $\phi(t)$ is called its *phase process*. The radial and phase processes of the standard 2-dimensional Ornstein-Uhlenbeck process are known to satisfy [31, 45, 46],

$$dZ(t) = \left[\frac{1}{2Z(t)} - Z(t) \right] dt + dW(t), \tag{3.7}$$

$$d\phi(t) = \frac{1}{Z(t)} db(t). \tag{3.8}$$

Here $W(t)$ is a standard Brownian motion, and $b(t)$ is an independent Brownian motion run on a unit circle. From (3.6) we see that $d\phi(t)$ is a stochastic process of phase 'slips' added to the regular phase progression $\mathbb{R}_{-\omega_d t}$. Further information follows from (3.7), (3.8) about the process of 'quasicycles,' as the process $\mathbb{V}(t)$ of oscillations sustained by noise is called.

The same modulus process, $Z(t)$, defined by Equation (3.7), describes the excursions away from zero of the Ornstein-Uhlenbeck process $S(t)$. We saw this process before as (2.36) in Section 2.5 where it was the process of subthreshold oscillations in a Morris-Lecar neuron model. It is itself nearly an O-U process, the term $1/2Z(t)$ in the drift coefficient serving to keep the process positive, once it has a positive start.

On the other hand, the factor $1/Z(t)$ in equation (3.8), which generates the phase process, has an essential role in explaining the burstiness of the gamma-frequency oscillations in the LFP. The factor $\mathbb{R}_{\omega_d}$ supplies the basic gamma frequency ω_d, the phase of which is spread by the Brownian noise $b(t)$ in the stochastic phase process $\phi(t)$. During time intervals when $Z(t)$ is away from zero, the spreading is small or moderate. But at times when the amplitude process $Z(t)$ is near zero, the phase slips are large, frequent, and stochastic, and the identity of the ω_d gamma frequency is completely lost. This causes the recorded gamma frequency oscillations to be bursty, or intermittent [3].

$\mathcal{PP}3.2.1$ The statistics of this phenomenon invite further analysis. By setting a small threshold of amplitude, $Z(t)$ a distribution of the lengths of excursions above that threshold can be defined and investigated in order to obtain distributional information about the durations of the gamma bursts.

The same structure holds wherever we find quasicycles, e.g, in neural models [3, 44], as well as in predator–prey and epidemic models [47, 48]. We can write any linear system of the form (3.2) directly in terms of its amplitude and phase processes. To identify these processes as in (3.7) and (3.8), however, one needs the approximation (3.6).

$\mathcal{PP}3.2.2$ So far this interesting identification has been exploited rather little in other contexts by probabilists or by others.

3.3 Wilson-Cowan population dynamics

In Section 3.2 we looked at a population model where we wrote the equations of the interactions of the excitatory and inhibitory populations as if each one were a unit, as in [3, 41]. One may wish, however, to begin with a particular model for the firing rates of individual neurons. In [44], individual neurons are binary, with active and quiescent states. The transition rates depend on the population state according to the neurons' inputs S_E, S_I. The firing rates, stochastic rates of transition from quiescent to active state, are

$$\beta_E f(S_E(t)) = \beta_E f\left(\frac{w_{EE}k}{N_E} - \frac{w_{EI}l}{N_I} + h_E\right) \tag{3.9}$$

$$\beta_I f(S_I(t)) = \beta_I f\left(\frac{w_{IE}k}{N_E} - \frac{w_{II}l}{N_I} + h_I\right) \tag{3.10}$$

for quiescent excitatory and inhibitory neurons, respectively, where f is the sigmoid function

$$f(s) = \frac{1}{1 + e^{-s}}. \tag{3.11}$$

The stochastic rates for transitions from the active to the quiescent state are constants α_i. The total input is

$$S_i(t) = \sum_j w_{ij}a_j(t) + h_j, \tag{3.12}$$

where w_{ij} are weights reflecting the strengths of the synaptic connections between neurons j and i, and $a_j(t) = 1$ if the jth neuron is active at time t and 0 otherwise. To model a sparse network we make many of the $w_{ij} = 0$. The Gillespie algorithm is used to simulate the process $(k_t, l_t) = $ numbers of active (excitatory, inhibitory) neurons at time t. Such simulations were not feasible when the Wilson-Cowan equations were proposed in 1972 [40]. Now it turns out that the linearization we have seen before in Section 3.2, called the 'linear noise approximation,' lets us write the Poisson-type model as a stochastic differential equation system if the noise is evaluated at the fixed point [44]. The linear noise approximation involves, for many but not infinitely many neurons of each type, either the van Kampen system size expansion (see [44]) or the Kurtz approximation [18], although it appears no one has as yet written out the latter method for this example.

$\mathcal{PP}$3.3.1 This looks like an attractive exercise for a probabilist.

For completeness, let us make a brief diversion to give more details about the linear noise approximation, as we will be using this idea in what follows as well. It is routine to center and linearize a nonlinear deterministic system about a fixed point, but it is not routine to do so for a corresponding nonlinear stochastic system. There are two equivalent ways in which such a linearization can be accomplished. Each applies to a wide class of models, in examples of which we are particularly interested, in which the deterministic part of a system of interacting stochastic differential equations generates a damped oscillation around a locally stable fixed point for some choice of parameter values.

Section 2 of [30] sketches the two methods by which any normed density-dependent Markov process, $X^N(t) = X(t)/N$ with values in $\mathbb{Z}^d$, for large N can be approximated by a diffusion process, with only small error. One way to do this is the method of Kurtz [18]. This method represents a $\mathbb{Z}^d$-valued Markov jump process as a sum of locally Poisson processes. In doing this each normed compensated, or conditionally centered, Poisson increment is replaced with a scaled Brownian increment. This introduces an error of order $(\log N)/N$. The resulting stochastic system is written as

$$d\tilde{X}^N(t) = F(\tilde{X}^N(t))dt + \frac{1}{\sqrt{N}}C(\tilde{X}^N(t))dW(t) \tag{3.13}$$

where F is the vector field of means of the terms in Kurtz's sum, and the $d \times d$ matrix function $C(z)$ is chosen so that $C(z)C(z)^* = B(z)$, which is the covariance function arising from interactions of the terms.

From Equation (3.13) we can identify the deterministic limit as $N \to \infty$ of the process $X(t)/N$ as the solution of

$$\frac{dv(t)}{dt} = F(v(t)) \tag{3.14}$$

The second term on the right-hand side of Equation (3.13) describes the fluctuations of $X(t)/N$ away from $v(t)$.These are of order $1/\sqrt{N}$. In certain parameter ranges the second term from Equation (3.13) generates sustained oscillations.

The method of van Kampen [49], which accomplishes the same end, is specified for a one-dimensional system, although it can be generalized to many dimensions. This method begins by stating the Ansatz

$$\frac{X(t)}{N} = v(t) + \frac{1}{\sqrt{N}}\xi(t). \tag{3.15}$$

In order to derive Equation (3.13) the Kolmogorov (or master) equation for the process $\xi(t)$ in (3.15) is expanded in powers of $1/\sqrt{N}$.

If the solution $v(t)$ of (3.14) has a single fixed point, z_{eq}, and $v(t) \to z_{eq}$ as $t \to \infty$ for all starting positions $v(0)$, we can linearize Equation (3.13)) to obtain the approximation for large N

$$\frac{X(t)}{N} \approx z_{eq} + \frac{1}{\sqrt{N}}\xi(t), \tag{3.16}$$

where the process $\xi(t)$ satisfies

$$d\xi(t) = -A\xi(t) + BdW(t), \tag{3.17}$$

and where $-A$ is the Jacobian of F evaluated at z_{eq}, and B is $B(z_{eq})$. The system (3.17) is termed the 'linearization of the stochastic system' or the 'linear noise approximation.' The process $\xi(t)/\sqrt{N}$ is the process of stochastic perturbations of order $1/\sqrt{N}$ of the system $X(t)/N$ about the fixed point z_{eq}.

Now back to our main narrative. The transition rates (3.9), (3.10) include the sigmoidal response function as in the original Wilson-Cowan model, whereas those of (3.1) do not. In fact if our attention is on the dynamics near the fixed point of the system as in (3.1) this does not make a noticeable difference. The inclusion of f produces a limit cycle in the parameter range where λ is positive and the fixed point is unstable, and this noisy limit cycle (Figure 3.3) is the focus of much attention, e.g., in [44]. In the nearby parameter range where λ is slightly negative and the fixed point is stable, there are sustained oscillations around it (Figure 3.2), and the analysis of Section 3.2 applies.

$\mathcal{PP}$3.3.2 A problem for research is to show how the power spectral density (the Fourier transform of the autocovariance function) of the model varies as we move

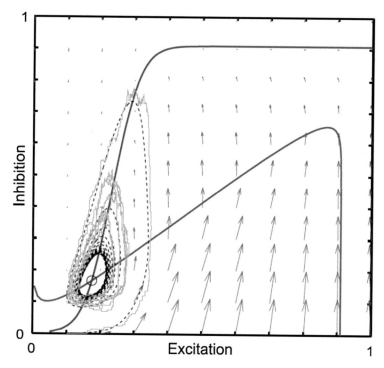

Fig. 3.3 Nullclines and limit cycle phase plane plots of the stochastic Wilson-Cowan population model for $\lambda > 0$. The fixed point is unstable. Here the paths start outside the limit cycle [44]. The limit cycle occurs in a different parameter range from that of the quasicycles in Figure 3.2. Reprinted with permission from [44]

parameters so that λ goes from positive to negative. We expect continuity. This is another example of smoothing by stochasticity that has not been investigated by stochastic analysis.

$\mathcal{PP}$3.3.3 A statistical problem is to devise a method to determine, from LFP data, whether oscillations arise from noisy limit cycles as in Figure 3.3 or from quasicycles as in Figure 3.2.

In [44] the individual neurons in the system are binary. They could have been modelled in any of the ways we saw in Chapter 2. Would that make a difference to the resulting population model? We think that the answer is 'no'; in other words, we suppose that the stochastic Wilson-Cowan population dynamics are dominant over the details of the individual neuron dynamics. But that question has not been examined carefully.

$\mathcal{PP}$3.3.4 One should be able to prove that when populations of neurons with any 'reasonable' single neuron dynamics are connected as in Figure 3.1, then a Wilson-Cowan-like model will govern the local field potential dynamics of the population.

3.4 Oscillations in an inhibitory network with delay

The approach of Brunel and Hakim [4] to the dynamics of stochastic neuron populations contrasts to that of Wilson and Cowan [40]. In [4], based on LIF neurons, the response to an inhibitory synaptic current is modelled with an explicit delay that appears to play the role of the E-I interactions of the Wilson-Cowan equations. In fact the delay shows up in the Fokker-Planck equation of [4], and is associated with oscillatory collective behavior of the system, which also, like [40], relies very much on inhibition. The results in [4], as in [40], include a Hopf bifurcation in the firing rate around the mean, with a growing fluctuation on one side of the bifurcation point and damping on the other.

An idea here is that perhaps the dominant eigenvalue is the essential determinant of results in both cases, especially in view of the smoothing effect of stochasticity. Alternate modelling by delay or by linking dynamics of two components appears to produce similar results in other stochastic contexts.

$\mathcal{PP}$3.4.1 Here lies an interesting problem. Can we show, using stochastic analysis, that the two approaches are nearly equivalent by some relevant measure?

A different approach to the problem of modelling oscillating neural networks with delay is that of Doiron, Lindner, Longtin, and colleagues [50, 51]. They studied a network of N uncoupled, noise-driven, spiking LIF neurons that receive a common inhibitory feedback based on the spiking output of all of them. Each LIF neuron in the network evolves according to

$$\frac{dv_i(t)}{dt} = -v_i(t) + \mu + \xi_i(t) + \sqrt{1-c}\,\eta_i(t) + \sqrt{c}\,\eta_c(t) + f(t), \qquad (3.18)$$

where μ is a common, constant current input, $\eta_i(t)$ is (i.i.d) intrinsic noise, $\eta_c(t)$ is extrinsic noise, c is the correlation of the external noise across the LIF neurons, and $f(t)$ is delayed inhibitory feedback based on the convolved spike train of all of the LIF neurons. When $c = 0$ the neurons each receive independent (internal only) noise, and when $c = 1$ they all receive the same external noise. The inhibitory feedback takes the form

$$f(t) = \frac{G}{N} \int_{\tau_D}^{\infty} \frac{\tau - \tau_D}{\tau_S^2} \, exp\left[-\frac{\tau - \tau_D}{\tau_S}\right] \sum_{i=1}^{N} y_i(t - \tau)\, d\tau, \qquad (3.19)$$

where $G < 0$ is the strength of the common feedback (i.e., it is inhibitory), τ_D is the transmission delay of the feedback, τ_S is a decay time related to synaptic transmission time, and $y_i(t) = \sum_j \delta(t - t_{i,j})$ is the output spike train of the ith LIF neuron at the jth instants of threshold crossing.

Using Fourier transforms of the zero-average output spike trains, they computed the power spectrum of the composite network output spike train, as well as that of individual neurons in the network. These power spectra show a peak at a particular frequency (i.e., an oscillation) when the noise inputs driving the neurons are correlated. The frequency of the peak depends on the amount of correlation

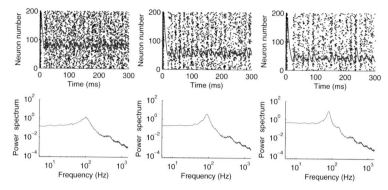

Fig. 3.4 Oscillatory behavior of population of neuron firings. Top: raster plot in which each dot represents a spike from the numbered neuron (vertical axis) at time t (horizontal axis). Red dots mark firings of neuron number 100. Superimposed blue traces represent the oscillating local field potential (LFP) generated by the neurons, all taken together, at each time point. Bottom: Power spectral density with a peak at the frequency that is obvious in the LFP trace. Notice that the power spectrum resembles that of a standard O-U process except for the peak; it is flat for the lower frequencies and above a particular frequency is like $1/f^2$. Reprinted with permission from [52]

and also on delay from a composite inhibitory feedback. Interestingly, the power spectra do not resemble the Lorantzians-with-a-peak that arise in the rate models we discussed in Section 3.3 and that also occur in the next model with delay that we discuss here (see [3] and Figure 3.4 bottom for examples). Instead the spectra generally increase with frequency up to and after the peak. The theory accounts very well for the response of lateral line neurons of electric fish to certain kinds of stimulus inputs [50].

$\mathcal{PP}$3.4.2 Does the difference in power spectra just noted point to an inconsistency between rate and spiking models in general or does the difference arise from the particulars of the neural networks modelled? Can the expressions for power spectra in the two theories be reconciled?

Now we look at yet another network model with delay where each binary neuron has two states, quiescent or spiking, or in the terminology of Dumont et al. [52] quiescent and active. This model has a finite number of interconnected interneurons; again, as in [4], they are all inhibitory. The transition probabilities for each neuron are

$$P(active \rightarrow quiescent, during\ dt) = \alpha\ dt$$
$$P(quiescent \rightarrow active, during\ dt) = \beta f(s)ds \tag{3.20}$$

where f is, again, the response function $f(s) = 1/(1 + e^{-s})$, and $s(t) = h - \omega$ $l(t - \tau)/N$. Here h comes from an input external to the network, $l(t)$ is the number of active neurons at time t, and τ is an axonal conduction delay. So $l(t - \tau)/N$ is the fraction of active neurons at time $t - \tau$ and their common inhibition arrives at the neuron in our focus at time t. The minus sign in the expression for $s(t)$ reflects this inhibition. Figure 3.4 shows the result of numerical simulations of this 'spiking' system. The oscillatory frequency that appears, both in the raster plot and in the power spectral density (see figure caption), is produced by the delay, and depends on the coefficient τ, which is taken to be 1. The ISI histogram in [52] (not shown here) tends to be multimodal by reason of stochastic facilitation as in Figure 1.1.

$\mathcal{PP}$3.4.3 A full stochastic analysis of this effect, together with the effects shown in Figure 3.4, is lacking, although analytic expressions for the power spectra are available [52].

Dumont et al. [52] supply a corresponding stochastic rate model that is written in terms of the fraction of active neurons at time t, $r(t) = l(t)/N$,

$$dr(t) = \left[-\alpha r(t) + [1 - r(t)]\beta f[s(t)]\right] dt + \frac{1}{\sqrt{N}} \sqrt{\alpha r(t) + [1 - r(t)]\beta f[(s(t)]}\, d\eta(t),$$
$$(3.21)$$

where η is a Brownian motion.

From the van Kampen system size expansion or again from [18], we have the linear noise approximation

$$r(t) = r_0(t) + \frac{1}{\sqrt{N}}\xi(t).$$
$$(3.22)$$

Here $r_0(t)$ satisfies the deterministic equation

$$\frac{d\, r_0(t)}{dt} = -\alpha r_0(t) + \left[1 - r_0(t)\right]\beta f\left[s_0(t)\right],$$
$$(3.23)$$

where $s_0(t) = h - \omega r_0(t - \tau)$, and $\xi(t)$ satisfies the stochastic delay equation

$$d\xi(t) = \left[\alpha\xi(t) - \beta f[s_0(t)]\xi(t) - [1 - r_0(t)]\beta f'[s_0(t)]\omega\xi(t - \tau)\right] dt$$

$$+ \sqrt{\alpha r_0(t) + [1 - r_0(t)]\beta f[s_0(t)]}\, d\eta(t).$$
$$(3.24)$$

The power spectral density can now be computed, and looks very much like Figure 3.4, bottom row.

An interesting point is that the parameters have been chosen so that the real parts of the eigenvalues are negative in the first two parts of Figure 3.4, and a bit positive in the last part. Hence we may be seeing the histogram of damped oscillations sustained by noise in the first two parts of Figure 3.4, and that of a noisy limit cycle in the last part. A consequence of the sigmoidal response function in this delay model is that a positive real part of the dominant eigenvalue gives us a noisy limit cycle rather than divergence. The fact that the histogram appears to change continuously and slowly across this bifurcation of the deterministic model suggests that the effect of smoothing by noise is present.

$\mathcal{PP}$3.4.4 It is an inviting problem to show this analytically.

$\mathcal{PP}$3.4.5 An additional problem is to prove an analogue of the approximation of [30] for the case of damped oscillations produced by a delay equation, and converted to sustained oscillations by noise. Such a result was obtained using multiscale analysis in [53]. The solution of the stochastic delay equation can be approximated, we expect, by a multiple of a rotation times a delay-Ornstein-Uhlenbeck process, when the real part of the dominant eigenvalue is near 0.

$\mathcal{PP}$3.4.6 Finally, the remark about the possibility of relating the Brunel-Hakim approach [4] to a Wilson-Cowan-type system, like (3.1), through dominant eigenvalues applies also to the Dumont et al. [52] approach.

3.5 Kuramoto synchronization of quasicycles

Bursty oscillations in EEG or local field potential data result from the interaction of subpopulations of neurons in a chunk of neural tissue. An influential model of one way in which such interactions can result in synchronized population oscillation dynamics was suggested by Kuramoto [54]. The Kuramoto model [54, 55], which describes how phases of coupled oscillators evolve in time, has been studied in many contexts, including populations of neurons, and usually for oscillators that are deterministic except for having a narrow distribution of natural frequencies. The typical result, which has been proved in several scenarios [54–56], is that synchronization increases with coupling strength, tending to a discontinuity at a critical coupling value as the number of oscillators increases.

$\mathcal{PP}$3.5.1 Here we describe a numerical result [57] in the stochastic context of quasicycles, which needs a theoretical proof.

Suppose, as in Section 3.2, we have a system of N Wilson-Cowan-type models

$$\tau_{Ei}dV_{Ei}(t) = (-V_{Ei}(t) + S_{EEi}V_{Ei}(t) - S_{EIi}V_{Ii}(t))dt + \sigma_{Ei}dW_{Ei}(t)$$

$$\tau_{Ii}dV_{Ii}(t) = (-V_{Ii}(t) - S_{IIi}V_{Ii}(t) + S_{IEi}V_{Ei}(t))dt + \sigma_{Ii}dW_{Ii}(t) \qquad (3.25)$$

where $i = 1, \ldots, N$. We have the approximation

$$\mathbb{V}_i(t) \approx \mathbb{V}_i^*(t) = \frac{\sigma_i}{\sqrt{\lambda_i}} \mathbb{Q}_i |\mathbb{S}_i(\lambda_i t)| [\cos(-\omega_{di}t + \phi(\lambda_i t)), \sin(-\omega_{di}t + \phi(\lambda_i t))]^\top,$$
$$(3.26)$$

where $|\mathbb{S}_i(\lambda_i t)| = \sqrt{S_{1i}(\lambda_i t)^2 + S_{2i}(\lambda_i t)^2}$ and $\theta_i(t) = -\omega_{di}(t) + \phi_i(\lambda_i t)$ are the amplitude and phase processes of $\mathbb{V}_i^*(t)$, which satisfy

$$dZ_i(t) = \left[\frac{1}{2Z_i(t)} - Z_i(t) \right] dt + dW_i(t), \qquad (3.27)$$

$$d\phi_i(t) = \frac{1}{Z_i(t)} db_i(t). \qquad (3.28)$$

Suppose we couple the phase and amplitude processes according to

$$d\theta_i(t) = \left(-\omega_{di} + \frac{1}{2N} \sum_{j=1}^{N} \frac{Z_j(\lambda_j t)}{Z_i(\lambda_i t)} \mathbb{C}_{ij} \sin(\theta_j(t) - \theta_i(t)) \right) dt + \frac{db(\lambda_i t)}{Z_i(\lambda_i t)}, \qquad (3.29)$$

and

$$dZ_i(\lambda_i t) = \left[\frac{\sigma_i}{\sqrt{\lambda_i}} ||\mathbb{Q}_i|| \left(\frac{1}{2Z_i(\lambda_i t)} - Z_i(\lambda_i t) \right) \right] dt$$

$$+ \left[\frac{1}{2N} \sum_{j=1}^{N} \mathbb{C}_{ij}(Z_j(\lambda_j t) - Z_i(\lambda_i t)) \right] dt + \frac{\sigma_i}{\sqrt{\lambda_i}} ||\mathbb{Q}_i|| dW_t(\lambda_i t). \qquad (3.30)$$

The matrix $\mathbb{C}$ specifies the degree of coupling between each pair of phase processes, θ_i and θ_j, and between the amplitude processes Z_i and Z_j. The factor Z_j/Z_i, which appears in the coupling of phases, reconciles the fact that subpopulations i and j have different amplitude processes.

The 'phase locking index' is usually used as a measure of phase synchronization. Imagine a collection of points $e^{i\theta_j}$ moving about on the unit circle in the complex plane. The quantity

$$\rho e^{i\psi} = \frac{1}{N} \sum_{j=1}^{N} e^{i\theta_j} \qquad (3.31)$$

is the centroid of the phases, θ_j, and the radius, $\rho(t)$, the phase locking index, measures phase coherence (Figure 3.5).

Simulations for the case where all C_{ij} are equal, in [57], show that $\rho = \rho(t)$, for t large enough to assure stationarity, is generally increasing with $||C||$, the coupling strength, and tends toward critical behavior for large N (Figure 3.6).

Fig. 3.5 Phase locking index is the length of the vector ρ. Red dots represent the phases of several different E-I processes at a given time point

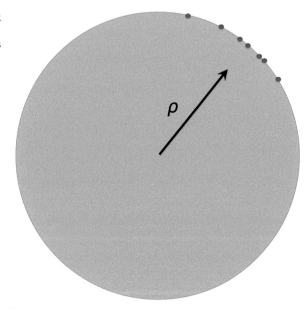

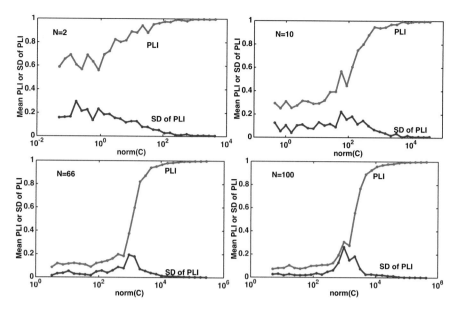

Fig. 3.6 Synchronization as a function of coupling in a stochastic Kuramoto model. The phase locking index (PLI or ρ) across all of the E-I processes increases with increasing coupling strength for various numbers of processes, reaching an asymptote at $\rho = 1$. ρ increases very rapidly near a critical coupling, as in the original Kuramoto system. Notice that the standard deviation (SD) of the values of ρ across the 10 realizations represented by each graph is generally rather small except near the critical coupling, where it is quite large. This might represent a 'chimeric' route to synchronization in this model. Reprinted with permission from [57]

$\mathcal{PP}$3.5.2 The immediate problem here is to prove that in the large N limit there is a critical value of $||C||$.

$\mathcal{PP}$3.5.3 An interesting question is whether, and if so where, does spatial structure, i.e. the details of an inhomogeneous matrix $\mathbb{C}$, begin to matter for synchronization?

$\mathcal{PP}$3.5.4 This question could be posed in terms of the connectivity of random graphs.

It would seem that questions and results about synchronization with Kuramoto-type coupling translate to the setting of dynamic graph theoretical models, where synaptic strength, C_{ij}, becomes probabilities of connections between nodes. Such a graphical model appearing in [58] is described in Section 4.6.

Chapter 4
Spatially Structured Neural Systems

Stochastic processes on a lattice, or on a network, are popular topics in probability currently. Much of the mathematical work in neuroscience, as indicated in previous chapters, has focussed on local phenomena. The models are written in terms of systems of ODEs and corresponding SDEs, as in previous sections, whether we are thinking of single neurons or local effects involving many neurons. There have been some recent suggestions, though, that we could frame spatial interactions of neurons or sub-populations of neurons in terms of reaction–diffusions or corresponding systems of stochastic partial differential equations. These could combine a reaction process in time, perhaps a firing rate model, locally, with copies of the local model positioned at nodes of a lattice or network, and a spatial interaction, also in time. The structure of the spatial interaction need not be homogeneous, as indeed neural networks are often not. The lattice, or network, might also incorporate various spatial and temporal scales, from interactions of individual neurons, to interactions of nodes consisting of from a few to many neurons, and might extend to interactions of brain regions encompassing an entire brain. McKane et al. [43] indicate how such a structure can be described in a notation that leads to familiar stochastic objects.

4.1 Reaction–diffusion equations

A simple reaction equation has the form $dV/dt = f(V_t, t)$, where V_t is a finite (e.g., 2-) dimensional function such as (V_{Et}, V_{It}) the voltages at time t of an excitatory neuron paired with an inhibitory neuron. If we have a 2-dimensional spatial array of neuron pairs, where the pairs interact with neighboring or nearby pairs, the interaction may be modelled as a discrete Laplacian. In this case the deterministic differential scheme may be thought of formally as

$$\partial V(x,t)/\partial t = D\nabla_x^2 V(x,t) + f(V(x,t)) \tag{4.1}$$

© Springer International Publishing Switzerland 2016
P.E. Greenwood, L.M. Ward, *Stochastic Neuron Models*, Mathematical Biosciences
Institute Lecture Series, DOI 10.1007/978-3-319-26911-5_4

This is a pair of reaction–diffusion equations where the diffusion term is the first term on the right and the reaction term is the second term. The terminology comes from chemistry. We can think of the discretized Laplacian as a local coupling.

Now we come to the question: how should noise be modelled? If we go back to the simple reaction equation, we recall that noise is modelled by interpreting the (deterministic) rates appearing in the function $f(V_t, t)$ as stochastic rates, i.e., if the rates are Poisson parameters, λ, then the probability of an event, y, in time dt is λdt. The typical treatment writes the stochastic transition rates for various states defining a Markov chain. If the states are numbers of individuals (neurons) in a category (could be excited or quiet as in Section 3.1), one divides by the total population, say N. Then either one of two methods leads to a stochastic system of equations. One can apply the Kurtz approximation [18], Section 3.3, to see that the stochastic process (Markov chain) is closely approximated, for large N, by a system of the form

$$dV(t)/dt = V^* + \frac{1}{\sqrt{N}}\xi(t), \tag{4.2}$$

where V^* is a solution of the deterministic system, and $\xi(t)/N$ is a diffusion that approximates the departure of the process from the deterministic solution. One can arrive at the same approximation by writing the 'master equation,' or Kolmogorov equation, of the Markov chain and applying a van Kampen, or system size, expansion [49], as we described in Section 3.3. We will see that this same approximation can be obtained in the case of a reaction–diffusion system.

To understand the possible relevance of stochastic reaction–diffusion equations for, e.g., visual phenomena, we review the concept of Turing instability. Turing [59] pointed out that a small perturbation of a stationary homogeneous state in a reaction–diffusion system can yield stable non-homogeneous spatial patterns based on certain wave numbers in space. These patterns are called Turing instabilities and they form only when the coefficients of the two Laplacian terms in the reaction–diffusion are sufficiently different. This corresponds to the diffusion motion of one species being predominant in the case of predator–prey systems, or one chemical diffusing more rapidly than another in the case of chemical systems. In the case of neural systems interactions replace actual movement and the asymmetry could be less. It has been shown, however, that in the case of stochastic reaction–diffusion systems the parameter range under which patterns form and persist is generally broader than the deterministic case [43]. Time-evolving patterns that would be damped in a deterministic model are sustained in the presence of stochasticity. This corresponds to the phenomenon that quasicycles, in stochastic dynamical systems indexed by time, produce effective bistability in a region far exceeding the deterministic bistability region. We have seen this in Section 2.5 and in Figure 2.10. Patterns that are 'driven by' or 'sustained by' noise are called stochastic patterns or 'quasipatterns' [60, 61, 48].

In applying these ideas to vision, we note that in the normal state patterns of excitation on the visual cortex V1 are driven by inputs from the retina (via the lateral geniculate nucleus of the thalamus, a sophisticated relay station in the visual pathway) generated by sensory stimuli. Visual hallucinations may arise when internally generated spontaneous (Turing) patterns of neural excitation overwhelm external-stimulus-generated input. Alternatively, it may be that in the normal state vision consists of excitation generated by external stimuli interacting with internally generated patterns of excitation. In either case it is of interest to understand from models how spatial patterns in vision and other sensory systems may be generated internally in the nervous system [62, 63, 64, 65].

We will see that for a parameter set for which the deterministic reaction–diffusion model has no 'unstable' behavior, but only the uniform stationary state, the stochastic model can display interesting signs of pattern, in terms of computed power spectral density, and also in terms of power spectral density based on simulations. This effect seems to be a space-time version of the noise-sustained oscillations we saw in purely 'reactive' systems, i.e., SDEs, for instance the Morris-Lecar single neuron model, Section 2.5 and in the Wilson-Cowan population model, Section 3.2.

4.2 The Kolmogorov equation for a stochastic reaction–diffusion system

A useful notation has been developed by McKane et al. [43], who wrote in terms of types of constituents with numbers $l, m, \ldots$ at a given time. Discrete spatial structure is introduced by locating types in a spatial domain or small volume, or it could be a graphical node, indexed by $j = 1, 2, \ldots$ at a given time. Here we will say 'neurons' often meaning 'local families of neurons' as in Chapter 3. We will think of the types, or 'species,' as types of neurons together with their states, say binary, and the spatial structures as a 2-dimensional rectangular grid with mesh 1, so that we can follow the notation of [43]. We use a combined index $J = (j, y)$ for the spatial and type indexes. Let $n_J, J = (j, y)$, denote the number of constituents of a particular domain, j, of a particular type, y, and $\mathbf{n}$ the vector of these numbers, which specifies the state of the system. The transition rate from state $\mathbf{n}$ to state $\mathbf{n}'$ is denoted by $\mathbf{T}(\mathbf{n}'|\mathbf{n})$. These will be defined as the same stochastic rates that appear in the 'reaction' part of the dynamics when the change from $\mathbf{n}$ to $\mathbf{n}'$ is only in the indices referring to type. Transition rates from $\mathbf{n}$ to $\mathbf{n}'$ when the change refers to location will be described later in this section.

The probability of finding the system in state $\mathbf{n}$ at time t, $P_\mathbf{n}(t)$, satisfies the Kolmogorov equation

$$\frac{dP_\mathbf{n}(t)}{dt} = \sum_{\mathbf{n}' \neq \mathbf{n}} T(\mathbf{n}|\mathbf{n}')P_{\mathbf{n}'}(t) - \sum_{\mathbf{n}' \neq \mathbf{n}} T(\mathbf{n}|\mathbf{n}')P_\mathbf{n}(t). \tag{4.3}$$

The somewhat awkward notation allows this simple form. The sum in (4.3) can be restricted to include only those transitions that actually take place with positive rates, usually a small subset. Suppose there are L different species of neurons in the system, labelled by type and domain (location). Now we number them and denote them by $Z_I, I = 1, \ldots, L$. At a given time there will be n_I of the Z_Is, and the state of the system can be written as $\mathbf{n} = (n_1, \ldots, n_L)$. Any transition of our Markov process can be written as

$$\sum_{I=1}^{L} r_{I\mu} Z_I \rightarrow \sum_{I=1}^{L} p_{I\mu} Z_I, \quad \mu = 1, \ldots, M, \tag{4.4}$$

where $r_{I\mu}, p_{I\mu}, I = 1, \ldots, L; \mu = 1, \ldots, M$ are the numbers of input and output neurons of species (type, domain) involved in a transition of type μ. There are M possible types of transition μ. Most of the $r_{I\mu}, p_{I\mu}$ will be zero, but the notation is very useful. The matrix

$$\nu_{I\mu} \equiv r_{I\mu} - p_{I\mu} \tag{4.5}$$

gives the numbers of neurons of species Z_I transformed by transition μ. In the notation of formula (2) of [43] we can write $\mathbf{n}' = \mathbf{n} - \boldsymbol{\nu}$ where $\boldsymbol{\nu}_\mu = (\nu_{1\mu}, \ldots, \nu_{L\mu})$ and the Kolmogorov equation takes the form

$$\frac{dP_{\mathbf{n}}(t)}{dt} = \sum_{\mu=1}^{M} \left[T_\mu(\mathbf{n}|\mathbf{n} - \boldsymbol{\nu}_\mu) P_{\mathbf{n}-\mu}(t) - T_\mu(\mathbf{n} + \boldsymbol{\nu}_\mu|\mathbf{n}) P_{\mathbf{n}}(t) \right]. \tag{4.6}$$

With this and some additional notation we will be able to write the system of stochastic differential equations that is the Kurtz [18] approximation to the density dependent Markov process at the microscopic scale, defined via the transition rates $T(\mathbf{n}'|\mathbf{n})$. First we write the differential system for the limit as V, the total 'volume' of the system, goes to infinity.

From the Kolmogorov equation (4.3) we see that the mean of $\mathbf{n}(t)$ satisfies

$$\frac{d\langle \mathbf{n}(t) \rangle}{dt} = \sum_{\mu=1}^{M} \boldsymbol{\nu}_\mu \langle T_\mu(\mathbf{n} + \boldsymbol{\nu}_\mu|\mathbf{n}) \rangle. \tag{4.7}$$

In [43] a limit process is constructed formally. We adumbrate this construction in the following section.

$\mathcal{PP}$4.2.1 A solid stochastic footing for this limit process would be desirable.

4.3   The macroscopic and mesoscopic models

Let us introduce new, scaled, density variables

$$y_I = \lim_{V \to \infty} \frac{\langle n_I \rangle}{V}, \quad I = 1. \ldots, L, \tag{4.8}$$

assuming these limits exist and are finite. We let

$$f_\mu(\mathbf{y}) = \lim_{V \to \infty} \langle T_\mu(\mathbf{n} + \mathbf{v}_\mu | \mathbf{n}) \rangle. \tag{4.9}$$

The system of equations

$$\frac{dy_I}{dt} = A_I(\mathbf{y}) \equiv \sum_{\mu=1}^{M} v_{I\mu} f_\mu(\mathbf{y}), \quad I = 1, \ldots, L, \tag{4.10}$$

defines the deterministic (macroscopic) system that is the limit of our Markov (microscopic) process as $V \to \infty$. The Kurtz approximation [18] or equivalently the van Kampen expansion approximation (see exposition in Section 3.3) is obtained by taking V large but *not* going to infinity. Using the new, scaled, density variable in equation (4.8) and replacing $T_\mu(\mathbf{n} + \mathbf{v}_\mu | \mathbf{n})$ by $f_\mu(\mathbf{y})$ as in (4.9), we rewrite the Kolmogorov equation (4.6) as

$$\frac{\partial P(\mathbf{y}, t)}{\partial t} = \sum_{\mu=1}^{M} \left[f_\mu \left(\mathbf{y} - \frac{\mathbf{v}_\mu}{V} \right) P \left(\mathbf{y} - \frac{\mathbf{v}_\mu}{V}, t \right) - f_\mu(\mathbf{y}) P(\mathbf{y}, t) \right]. \tag{4.11}$$

Now expand the functions P and f as Taylor series around $\mathbf{y}$ and truncate at order $O(V^{-2})$ to obtain a Kolmogorov equation for the stochastic reaction–diffusion system,

$$\frac{\partial P(\mathbf{y}, t)}{\partial t} = -\sum \frac{\partial}{\partial y_I} \left[A_I(\mathbf{y}) P(\mathbf{y}, t) \right] + \frac{1}{2V} \sum_{I,J} \frac{\partial^2}{\partial y_I \partial y_J} \left[B_{IJ}(\mathbf{y}) P(\mathbf{y}, t) \right], \tag{4.12}$$

where $A_I(\mathbf{y})$ is defined by equation (4.10), and

$$B_{IJ}(\mathbf{y}) = \sum_{\mu=1}^{M} v_{I\mu} v_{J\mu} f_\mu(\mathbf{y}), \quad I, J = 1, \ldots, L,$$

which, as a matrix, is positive semi-definite, as probabilists know. The corresponding system of stochastic equations, which we could also obtain from the approximation result of Kurtz [18], is

$$dy_I = A_I(\mathbf{y})dt + \frac{1}{\sqrt{V}} \sum_J g_{IJ}(\mathbf{y})\eta_J(t), \quad I = 1, \ldots, L, \tag{4.13}$$

where the η_J are Brownian increments with

$$\langle \eta_I(t)\eta_J(t') \rangle = \delta_{IJ}\delta(t - t') \tag{4.14}$$

and

$$B_{IJ}(\mathbf{y}) = \sum_K g_{IK}(\mathbf{y})g_{JK}(\mathbf{y}). \tag{4.15}$$

An example of the system (4.13), where the noise term appears only in the second equation of a FitzHugh-Nagumo system, is (1) of [65]:

$$\epsilon \frac{du(t, n)}{dt} = u - \frac{u^3}{3} - w + \gamma \sum_{n'} \frac{1}{2kl^2}\left[u(t, n') - u(t, n)\right]$$

$$\frac{dw(t, n)}{dt} = u + a(n) + \sqrt{\frac{2D}{\tau_w l^k}}\, \xi(t, n), \tag{4.16}$$

where $u(t, n), w(t, n)$ are variable values at time t at location n, fast and slow, respectively, because $\epsilon = \tau_u/\tau_w \ll 1$ where τ_u, τ_w are time constants of the u, w processes, $k = 1, 2$ is the dimensionality of the space with spacing between processes l, γ is the coupling strength, $a(n)$ is activation (a random number in a certain range), and ξ is zero mean Gaussian noise with amplitude D. The sum over neighbors represents the discrete Laplace operator controlling local interactions. With $D = 0$ the system quickly reaches equilibrium. With sufficiently large coupling strength and $D > 0$ spatial patterns emerge: for an intermediate level of noise synchronized firings occur broadly across the space, whereas for lower levels of noise traveling waves of firing appear and for higher levels of noise the firings are random across the entire space.

McKane et al. [43] call the system defined by (4.13) or equivalently (4.12) the *mesoscopic* model, because it operates on a scale between the deterministic macroscopic system (4.10), and the microscopic scale Markov chain defined by the transition probabilities $T(\mathbf{n}'|\mathbf{n})$. In fact the other models in this chapter defined by diffusion approximations are mesoscopic in the same sense. If (4.13) is put on a solid footing with A_I a linear operator, so that (4.13) is a high dimensional version of the 2-dimensional system (3.2), we will see that under some eigenvalue conditions an approximation like (3.6) is valid on a particular hyperplane. This will lead us to an analytic description of stochastic Turing patterns in Section 4.4.

$\mathcal{PP}$4.3.1 The last three sentences of this paragraph are optimistic speculation. This program has not been carried out for a spatial neural model.

4.4 Neuronal stochastic (quasi)patterns

There have been several suggestions (see, e.g., [61, 62, 66]), that stochastic Turing quasipatterns are relevant to V1 cortex. In particular, a detailed treatment can be found in the supplementary materials of [62]. There is a modelling choice at the point of specifying how adjacent neural nodes, in a lattice or in a more general network structure, interact. What is needed is a pattern of spatial interactions that mimics a discretized Laplacian. A reasonable choice seems to be a 'Mexican hat' pattern [67, 68], where at each spatial site the system is excited by nearby systems and inhibited by sites slightly further away. Suppose such a system is in place. The transition probabilities $T(\mathbf{n'}|\mathbf{n})$ in Section 4.2 pertaining to the discrete diffusion transitions need to be modelled and the parameters involved estimated. This has been done, formally, in the [43] treatment of stochastic patterns in the Brusselator model. McKane et al. [43] are able, starting from the 'linear noise approximation' with stochastic equation (4.13) to compute examples of the power spectral density of this system, which show that frequency peaks, Figure 4.1, and hence stochastic patterns, exist, in addition to deterministic Turing patterns.

We return to the combined index $J = \{j, y\}$ where j is a spatial and y a type index, and now we separate the indices. Use of a 2-dimensional square lattice structure with periodic boundary for the spatial nodes, or domains, allows use of Fourier transforms for the spatial part of the system. Otherwise, for a network, one can use the network transform method of [48]. The discrete spatial Fourier transform is written

$$\tilde{f}_k = l^D \sum_{j=1}^{\Omega} e^{-ilk \cdot j} f_j \quad \text{with} \quad f_j = l^{-D} \Omega^{-1} \sum_{k=1}^{\Omega} e^{ilk \cdot j} \tilde{f}_k, \tag{4.17}$$

where l is the lattice mesh, Ω is the number of lattice points, j is the 2-dimensional spatial index, and k is its Fourier conjugate. The macroscopic equations (4.10) can be written

$$\frac{du_i}{d\tau} = g_l(u_i, v_i) + \alpha \Delta u_i \quad i = 1, \dots, \Omega \tag{4.18}$$

and

$$\frac{dv_i}{d\tau} = g_m(u_i, v_i) + \beta \Delta v_i, \quad i = 1, \dots, \Omega, \tag{4.19}$$

where $u_i = l_i/V, v_i = m_i/V$. The u, v are local densities in space of neurons in active, inactive states. The symbol Δ denotes the discrete Laplace operator,

$$\Delta f_j = \frac{2}{Z} \sum_{j' \in \partial j} (f_j - f_{j'}), \tag{4.20}$$

where $j' \in \partial j$ means j' is a nearest neighbor of j on the lattice, and Z is the number of neighbors or 'coordination number of the lattice.' The spatial Fourier transform of Δ is [47]

$$\tilde{\Delta}_k = \cos(k_1 l) - 1 + \cos(k_2 l) - 1. \tag{4.21}$$

One could take the limit in (4.18) as the lattice mesh l goes to 0 to obtain a traditional reaction–diffusion system, but the theory is simpler for the lattice system (in [60] a quantum field theory version is presented).

The linear noise approximation (4.13) written in the notation of (4.18), (4.19) is

$$y_I(t) = \langle y_I(t) \rangle + \frac{\xi_I(t)}{\sqrt{V}}, \quad I = 1, \dots, L, \tag{4.22}$$

where $\langle y_I(t) \rangle$ satisfies the macroscopic equation (4.10) and $\xi_I(t)$ satisfies

$$\frac{d\xi_I}{dt} = \sum_J \mathcal{J}_{IJ}(\langle \mathbf{y} \rangle) \xi_J + \sum_J g_{IJ}(\langle \mathbf{y} \rangle) \eta_J(t). \tag{4.23}$$

One obtains (4.22), (4.23) either from a Kurtz [18] approximation or from a van Kampen expansion as described in Section 3.3. Here $\mathcal{J}$ is the Jacobian of the system, and g is related to B, the covariance, as in (4.15). Because we are interested in the local behavior of the system near a fixed point of the macroscopic equation, both can be evaluated at the fixed point, y^*, producing $\mathcal{J}^*, B^*$. The spatial Fourier transform of (4.23) then looks like

$$\frac{\partial \tilde{\xi}_\gamma (k, t)}{\partial t} = \sum_{\delta=1}^{2} \mathcal{J}_{\gamma\delta}^*(k) \tilde{\xi}_\delta (k, t) + \sum_{\delta=1}^{2} g_{\gamma\delta}^*(k) \tilde{\eta}_\delta (k, t). \tag{4.24}$$

The Jacobian $\mathcal{J}_{(k)}^*$, for each k, has eigenvalues $\lambda_\gamma(k)$, $\gamma = 1, 2$. If $\mathrm{Re}(\lambda_\gamma(k))$ is positive, the homogeneous state is unstable for the *deterministic* equation. Then the deterministic solutions u and v have a sinusoidal profile about their fixed point for wave number k, and this is a Turing pattern. In the stochastic case we take the temporal Fourier transform of (4.24) to obtain

$$\tilde{\xi}_\gamma (k, \omega) = \sum_{\delta, \sigma} \Phi_{\gamma\delta}^{-1}(k, \omega) g_{\delta\sigma}^*(k) \tilde{\eta}_\sigma (k, \omega), \tag{4.25}$$

Fig. 4.1 Power spectra for the Brusselator model. Top: computed from Equation (4.26). Bottom: average of 200 stochastic simulations. Reprinted with permission from [43]

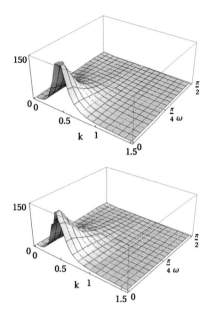

where $\Phi_{\gamma\delta}(k,\omega) = i\omega\delta_{\gamma\delta} - \mathcal{J}_{\gamma\delta}^*$, and the expression for the power spectrum is

$$P_\gamma(k,\omega) = \langle |\tilde{\xi}_\gamma(k,\omega)|^2 \rangle = \sum_{\delta\sigma} \Phi_{\gamma\delta}^{-1}(k,\omega)\tilde{B}_{\delta\sigma}^*(k)(\Phi^\dagger)_{\sigma\gamma}^{-1}(k,\omega). \qquad (4.26)$$

Figure 4.1 displays an example of this power spectrum both from calculation of (4.26) (top) and by simulation (bottom). We see that there are indeed stochastic patterns for the case of the Brusselator. There are also neural stochastic patterns. For example, in [65] stochastic patterns vary with noise level, with synchronization occurring at an optimal noise level.

Other work showing that stochastic Turing patterns exist and suggesting their use in modelling visual phenomena appears in [60, 62]. The power spectrum of [60], Figure 4.2, shows an example for a predator–prey model where there is a peak in ω (the 'reaction' coordinate of the power spectrum) for the wave number $k = 0$, and a (lower) peak in k for $\omega = 0$. This seems to mean that the wavelength of dominant quasipatterns is about the same as the mean-field value.

Damped deterministic patterns are 'sustained by noise' just as damped oscillations in ODE systems are sustained by noise in the corresponding SDE systems. If the power spectrum showed a peak at $k \neq 0, \omega \neq 0$, the pattern would have stochastically sustained traveling waves [66]. The groundwork has been laid for specific results about stochastic patterns in neural systems but the results are only suggestive and partial, using simulations.

$\mathcal{PP}$4.4.1 The challenge for probabilists here is modelling, computation, and analysis, all of it stochastic.

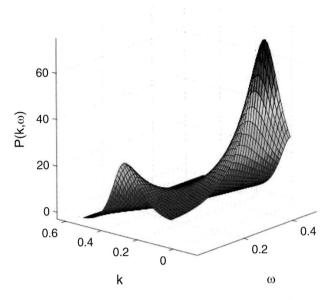

Fig. 4.2 Space-time power spectra indicating existence of both spatial and temporal stochastic patterns in a predator–prey model. Reprinted with permission from [60]

4.5 Synchronization in spatially structured systems

We discussed synchronization of quasicycles in spatially homogeneous population models in Section 3.5, and the existence of noise-driven spatial patterns in Sections 4.1, 4.2, 4.3, and 4.4. Now we turn to the idea that the detailed, and possibly inhomogeneous, spatial structure of a network might influence other aspects of its behavior, notably whether and how there is synchronization of individual neurons' firing patterns and other voltage oscillations that lead to a population dynamic such as that produced by the Kuramoto model. For spatially inhomogeneous networks, we describe first the structure itself rather than the stochastic process that operates on that structure.

One model of an inhomogeneous neural network is that of Izhikevich [20, 21]. In [20], 1000 Izhikevich neurons (see Section 2.3) are pulse-coupled. In pulse coupling each neuron receives inputs as constant-size spikes from the other neurons, multiplied by the relevant coupling strength, rather than receiving a continuous input current. Neurons also each receive i.i.d. voltage inputs during each simulated time interval. The total of 1000 neurons is divided between excitatory and inhibitory neurons in the proportions found in sensory cortex, roughly 4:1 for excitatory and inhibitory neurons respectively. Heterogeneity is introduced in two ways: first, by making a subset of the parameters of each neuron model (such as the parameters a, b, c, d, Section 2.3) random over a small interval around a desired value, and second, by making random the outgoing coupling strengths between the various

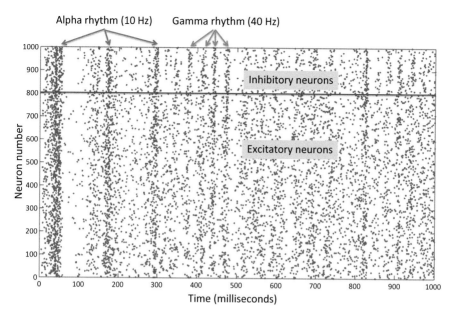

Fig. 4.3 Synchronization in the pulse-coupled (see text) Izhikevich model. Columns of dots in raster plot (dots represent neuron spikes) are neuron spikes occurring roughly at the same time, demonstrating spike synchronization at both alpha (8–14 Hz) and gamma (30–80 Hz) frequencies. Notice that although many neurons fire on some cycles of both rhythms, neuron firing is irregular and few, if any, neurons fire on every cycle. This is called 'sparse firing.' The raster plot represents one realization from running the Matlab program of [20]

neurons (uniform on the interval $[0, 0.5]$ for excitatory neurons, and uniform on the interval $[-1, 0]$ for inhibitory neurons). Izhikevich [20] demonstrated that his network's output exhibits spiking activity similar to that found in mammalian sensory cortex, including periods of spike synchronization at commonly observed frequencies (Figure 4.3), especially the behaviorly important ones of 10 Hz and 40 Hz. In Figure 4.3 synchronization is apparent from the dark vertical lines, indicating firing at very close time points, in a majority of the neurons represented in the raster plot. Synchronization is particularly prominent among the inhibitory neurons (see Section 3.4). The description of this system constitutes a model.

$\mathcal{PP}$4.5.1 A stochastic analysis might explain the firing patterns that are evident in Figure 4.3.

One question that has been raised prominently in the recent literature is whether particular local motifs, such as those shown for 3-neuron combinations in Figure 4.4, could have an influence on how neurons generate patterns of activity [69]. For example, it was found in [70] that gamma-band (30–50 Hz) power was not associated with stimulus input intensity in simulated networks that lacked reciprocal connections (like those in Figure 3.1), whereas these were significantly correlated in those networks that did contain sufficient reciprocal connections. They speculated

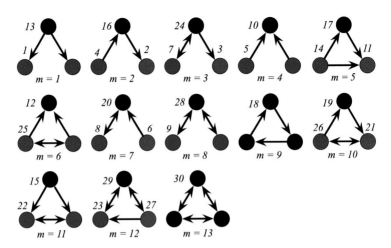

Fig. 4.4 All possible motifs and motif-roles for 3-neuron combinations. Arrows represent directions of connections with a synapse at the arrowhead. The important aspect of this figure for the present purpose is that there are 13 different ways (m, 'motifs') in which 3 neurons can be connected, and that these motifs might be important in generating temporal rhythms and spatial patterns. The neurons play a variety of different roles (numbered 1–30; called 'motif-roles') in forming the motifs. For example, the '13' next to the upper left neuron designates a 'two-source' role for that neuron. Reprinted with permission from [71]

that the distribution of motifs, like those in Figure 4.4, in the network would likely determine the extent to which neural network activity is associated with inputs.

$\mathcal{PP}$4.5.2 This seems to invite a stochastic model.

We suggest that the distribution and abundance of various types of motifs might indeed be important in networks with relatively few neurons, such as those found in, e.g., *C. elegans*, whose 'brain' has only roughly 300 neurons. In this organism, circuits of relatively few neurons control specific behaviors, such as mating, or reversal of locomotion after a shock to the substrate [39]. Thus, particular motifs are likely to participate in such control, and if they are not present, or even if a particular neuron is missing, the behavior cannot be carried out. In the much larger mammalian brain, however, it is possible that global network connectivity is more important than the local manner in which it is implemented.

In support of this contention, we offer an observation [72] based on a network similar to the Izhikevich network described above. We created such networks of Izhikevich neurons with various proportions of motifs, including several of the ones shown in Figure 4.4, as well as four- and five-element motifs. In all of these cases we found that the overall connectivity, defined as the size of a Tarjan component [73], or complete subgraph within the network, was critical in generating spiking synchronization, not how that component was locally constructed. Tarjan components of sufficient size, in this case at least several hundred Izhikevich neurons, including both excitatory and inhibitory types, generate sustained synchronized firing at a characteristic frequency, whereas sparsely connected networks without such a large complete component do not display such behavior. A caveat to this last statement is that appropriate levels of correlated noisy input can synchronize even unconnected neurons (see [74]).

Problems for probabilists abound here. Very little has been worked out theoretically regarding how the local spatial organization of a network of neurons affects its global activity, such as power spectrum, synchronization, and standing or traveling waves. According to the work of McKane et al. cited in Section 4.2, the spatial structure should be retained by the linear approximation process producing reaction–diffusion systems. But very little analysis has been applied to this problem when local spatial structure is inhomogeneous and effective system size varies over a wide range. It seems to us likely that results will be different depending on system size. Perhaps there is a critical system size where local details become unimportant, just as the details of individual neuron structure becomes unimportant at the population level.

$\mathcal{PP}$4.5.3 But the relevant stochastic analysis remains to be done.

4.6   Additional stochastic graph-based neuron models

In Section 4.1 we mentioned that discrete-space reaction–diffusion models can include local graphical or network weighted connections as the diffusion aspect. If we then drop the reaction part, we arrive at or near the area of interacting particle systems, including percolation theory on lattices or networks. We will assume that probabilists are familiar with the basics of percolation theory and may enjoy an introduction to what is called 'neuropercolation' [75]. In [75] a descriptive overview and orienting annotated bibliography, disjoint from our references to this point, is presented. Our challenge here is to bring these two lines of research closer together.

We begin by looking at [76], where local interactions of binary neurons are introduced using cellular automata formalism. Time is discrete. Our neurons sit at sites x in Z^2, take values $\alpha_x(t) = 0, 1$, and are influenced by their neighbors in $\Lambda(x)$, probably four neighbors in Z^2 plus the self, according to

$$P\big(\alpha_{t+1}(x) = 1 \,|\, \alpha_t(x) = 0\big) = \xi_1 \text{ if } \sum_{k \in \Lambda(x)} \alpha_k(t) \leq \frac{|\Lambda(x)|}{2}. \qquad (4.27)$$

In words, an inactive neuron at x is activated at time $t + 1$ if the majority of its neighbors are inactive at time t, with a small probability ξ_1, otherwise it is activated with probability $1 - \xi_1$, so the inactive neuron at x tends to follow its neighbors. Similarly, an active neuron at x tends to follow the majority of its neighbors, with ξ_2 replacing ξ_1.

There are some theoretical results for this model with ξ_1, ξ_2 approaching 0. The results say that the states 'all 0s' and 'all 1s' are metastable and it takes a very long time (with estimates) to move from one metastable state to the other [76].

A 'small world' model is constructed by introducing a few long connections into the neighborhood model. Simulations show the small world system moving abruptly from one metastable state to the other [77, 78].

Typical spatial patterns are plotted in [76] for various configurations of neighbors. These have similarities to patterns shown in [43] by a discrete reaction–diffusion system. It is not surprising, but also not completely understood, that there is a connection.

$\mathcal{PP}$4.6.1 There are several open problems here, including the characterization of additional structure produced by a reaction–diffusion type model that a diffusion model alone cannot reach.

Analytic results for an all-to-all network of discretized integrate-and-fire type neurons are obtained in [58]. The system has two attractors, characterized as synchrony and decoherence. A coupling parameter p measuring reliability of synapses decreases with system size N, so that $Np = \beta$. The parameter β has a critical point. In contrast to the classical Erdös-Renyi theory, here the critical β depends on a firing level k of the discretized voltage. An interesting aspect is that there is a middle regime of β where the large N stochastic system alternates between the two metastable states of synchrony and decoherence. The proof of convergence to a mean-field limit uses, among other things, results of [18]. The work of [58] and its follow-ups have been some of the main efforts to connect neural networks with random graph theory.

An additional aspect of spatially structured neural systems on a network is the reinforcement of neural connections, the development of effective synaptic weights. Most theories of learning involve repeated stimuli and rules about how systems evolve to respond appropriately. The reinforcement of neural connections can be modelled by what probabilists call an urn scheme [79], where an urn contains balls of various colors. A ball is drawn and replaced, perhaps, with another ball of the same color. If we identify colors with synapses, the reinforcement of a color corresponds to the reinforcement of a neural connection each time it is used. Problems about such an urn scheme with specific replacement rules are studied in [79]. Results are of the form: under certain conditions there are linearly stable equilibria to which the state of the urn scheme converges with positive probability [79].

$\mathcal{PP}$4.6.2 It is conjectured that such convergence happens with probability 1. A proof is lacking.

It is clear that such results transfer to the context of reinforcement of neural connections.

Chapter 5
The Bigger Picture

The most exciting current topic in neural modelling is no doubt at the whole-brain level. We want to understand how the nervous system 'works,' how we perceive, learn, remember, react appropriately, plan, decide, consider ourselves to be conscious. We are not there yet.

Several simplified large-scale spatiotemporal models of the entire brain, however, have been or are being developed and studied via simulations. The earliest such model is probably that of Nunez [80]. Nunez developed a 'cortical wave equation' that was used to explain EEG scalp potentials in terms of neural activity at synapses. In a more recent version of this work, Nunez [81] proposed a set of five second-order differential equations to describe brain-wide waves of cortical activity under conditions where nonlinearities are small and local and global resonances do not interact, each one, after space is averaged out, of the form:

$$\frac{d^2 Z}{dt^2} - \epsilon(1 - Z^2)\frac{dZ}{dt} + [k_n^2 - (1 + \epsilon)Z^2]Z = 0. \qquad (5.1)$$

In more complicated versions with larger nonlinearities and coupled resonances, the right-hand side is not zero. Equation (5.1) resembles that of the van der Pol relaxation oscillator [34], from which the FH-N equations were developed, where Z is 'synaptic action density,' similar to voltage, ϵ is a damping parameter, and $[k_n^2 - (1 + \epsilon)Z^2]Z$ is the restoring force. Numerical solutions to a more complicated set of partial differential equations, with space included, show temporal and spatial dynamics, comprised of multiple-frequency temporal oscillations and traveling and standing spatial waves in the brain. Section 4.4 describes initial analytical work in this direction.

Most recently, Nunez and Srinivassan [82] studied a generic global model of neural dynamics with both space and time included,

$$\rho G(\mathbf{r}, t) = 2\beta H(\mathbf{r}, t) - \alpha H(\mathbf{r}, t)^3, \qquad (5.2)$$

© Springer International Publishing Switzerland 2016
P.E. Greenwood, L.M. Ward, *Stochastic Neuron Models*, Mathematical Biosciences
Institute Lecture Series, DOI 10.1007/978-3-319-26911-5_5

where $G(\mathbf{r}, t)$ is the modulation of action potential density (similar to probability density of an action potential) around its background level, at location $\mathbf{r}$ (a 3-dimensional vector) in the brain at time t, and $H(\mathbf{r}, t)$ is the modulation of synaptic action density (again, like a probability density except now referring to synaptic transmission) around its background level at location $\mathbf{r}$ at time t. The constant ρ is the average number of excitatory synapses per axon connecting two cortical neurons, β is the background excitability controlled by chemical neuromodulators on long time scales, and α is the strength of inhibitory feedback.

The expression for $H(\mathbf{r}, t)$,

$$H(\mathbf{r}, t) = H_0(\mathbf{r}, t) + \int_0^\infty dv_1 \int_{cortex} \int \mathbb{R}(\mathbf{r}, \mathbf{r}_1, v_1) G\left(\mathbf{r}_1, t - \frac{|\mathbf{r} - \mathbf{r}_1|}{v_1}\right) d^2\mathbf{r}_1 \qquad (5.3)$$

is stochastic because of local sub-cortical input, $H_0(\mathbf{r}, t)$, which is assumed to be Poisson distributed spikes, and also an irregular action potential density, $G(\mathbf{r}, t)$, transported in by axons from other cortical areas and integrated across the same neocortical hemisphere. $\mathbb{R}(\mathbf{r}, \mathbf{r}_1, v_1)$ is the connectivity matrix whereby activity from neurons at location $\mathbf{r}_1$ affects that at location $\mathbf{r}$ at propagation velocity v_1.

Nunez and Srinivassan arrived at (5.2) by assuming a sigmoidal saturation relationship, similar to that employed by Wilson and Cowan [40], but here between action potential density and synaptic action density, which they expanded to obtain a quasi-linear model. In the linear version ($\alpha = 0$), a dynamic transfer function (or its inverse, a dispersion function) is obtained. A linear partial differential equation in $H(\mathbf{r}, t)$ can be derived from the dispersion function. One example of such an equation is

$$\frac{\partial^2 H}{\partial t^2} + 2v\gamma \frac{\partial H}{\partial t} + v^2\gamma^2 H - v^2 \frac{\partial^2 H}{\partial x^2} = \rho\left(v^2\gamma^2 + v\gamma\frac{\partial}{\partial t}\right)G + v^2\gamma^2 H_0, \qquad (5.4)$$

where x represents spatial location, v an average propagation velocity, and the other symbols are as above. Applying boundary conditions to this equation, based on (5.2), results in a nonlinear partial differential equation, and numerical and approximate solutions again yield dominant oscillatory modes in both space and time. In general, this system functions as a spatio-temporal filter with certain dominant frequencies, as in some of the models we have discussed above. Using additional assumptions about long range and local synaptic transmission times, Nunez and Srinivassan arrived at limit cycles and damped oscillations at specific frequencies, similar to the Wilson-Cowan model.

$\mathcal{PP}$5.1 Nunez and Srinivassan [82] neglected local neural activity in deriving their global model. A problem here is to relate more exactly their global model to the local models such as that of Wilson and Cowan and the others we have described above, including their stochastic aspects. A natural goal is to specify a model that links together at least three spatial scales in the brain: < 1 mm, several mms, and several cms. Note that the spatial scales also imply corresponding temporal scales.

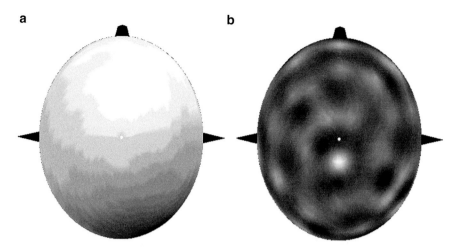

Fig. 5.1 Standing alpha frequency (8–13 Hz) waves in the brain. (a) Alpha potential amplitude on the scalp surface from a high-density EEG recording at the peak of the alpha wave. (b) Spatially high band pass filtered image of the same potentials based on an algorithm that estimates potential at the dura (a layer of tissue that separates the brain from the skull; see [82]). The relatively light and dark areas are approximately 180 deg out of phase with each other. Within each light or dark patch alpha oscillations are roughly in phase. Thus, those light and dark areas represent the maxima and minima of a spatial standing wave composed of local synchronized oscillations. Reprinted from [83] with permission

Nunez and Srinivassan [82] showed how the standing and traveling waves of global potential fields created by synaptic delays between relatively distant (at least several centimeters) brain regions, could have top-down effects on the activity of local (millimeter and smaller scales) information processing networks. The spatial frequency of the resulting waves depends strongly on overall cortical excitability, β as in (5.2). Figure 5.1b shows an example of such a global standing wave of alpha frequency oscillation in which a number of local alpha frequency rhythms, in different phases, are nested. The spatial frequency of the standing wave appears to be about 0.5 cycle per cm, or a wavelength of about 2 cm, based on an average human head of about 20 cm from front to back. Because these standing waves are composed of synaptic activity, they could enhance or interfere with local network activity, a top-down effect.

A somewhat more ambitious model by Izhikevich and Edelman [21] was first constructed as a model that incorporated 100 billion Izhikevich neurons of various types appropriate for their brain region and function, and subsequently scaled down for more extensive study. The model faithfully rendered the thalamocortical pathways (each area of neocortex connects directly to a particular area of the thalamus in the brainstem in a looping way) in a human brain to the resolution available from a diffusion tensor imaging (a way to derive neural pathways based on water transport direction deduced from MRI images) reconstruction of Izhikevich's own brain. The model was started in a spatially homogeneous state and run for a time period during which spike-timing-dependent-plasticity-induced changes in strength of synaptic connections created both short and long-lasting spatial patterns of oscillating neural activity.

At a spatial scale similar to that employed by Nunez and Srinivassan [82], the Virtual Brain project attempts to create realistic models of individual patients' brains in order to test new drug therapies, as well as to simulate scientific results from brain imaging, electroencephalography/magnetoencephalography (EEG/MEG), and intracranial studies (e.g., [84]). It is based on a generic stochastic differential equation that describes the activity of a network of connected neural populations $x_i, i = 1 \ldots N$ [85]:

$$\frac{d\Psi(x_i, t)}{dt} = N\Psi(x_i, t) + \int_\Gamma g_{local}(x_i, x_j) S[\Psi(x_j, t)] dx_j$$

$$+ \int_\Gamma g_{global}(x_k, x_l) S[\Psi(x_l, t - |x_k - x_l|/v)] dx_l + I(x_i, t) + \epsilon(x_i, t). \quad (5.5)$$

Here Γ is the entire network, or neural field, within which each x_i is located, $\Psi(x_i, t)$ is the activity of the entire network of nodes, x_i, over time, t, $g_{local}(x_i, x_j)$ is local, short-range, connectivity which is assumed to be instantaneous, $g_{global}(x_k, x_l)$ is long range connectivity which introduces a delay dependent on distance between the relevant loci in the neural field and transmission speed v between them, $I(x_i, t)$ is input, and $\epsilon(x_i, t)$ is stochastic noise. Local connectivity is typically spatially invariant and falls off with distance whereas global connectivity is highly heterogeneous. The model has been used to develop a forward model (a model that predicts potential on the surface of the head from electrical activity at every point inside the brain) for MEG [86] and used to give more insight into the relationship between neural activity and the functional magnetic resonance imaging (fMRI) blood oxygen level dependent (BOLD) signal [87].

Finally, there are several efforts underway to model the detailed anatomy of the human brain, including the BigBrain [88] from a joint Canadian-European consortium, and the Human Brain Atlas from the Allen Institute (http://human.brain-map.org). When such models are complete, inserting neuron models for the nodes, either individual neurons or small populations of them, will give us the opportunity of simulating the large-scale activity of the brain, including the stochasticity arising at smaller, faster scales. When faster computers are developed this may be possible to accomplish in real time. Thus we will in the future, at least from simulations, be able to gain an approximate understanding of how spatial and temporal patterns evolve in a realistic brain.

$\mathcal{PPPPP}$5.2 A very large problem for probabilists is to delve into the fundamental principles that govern such patterns, and, in particular, the role of stochasticity in brain function.

Competing interests

The authors declare that they have no competing interests.

Author's contributions

All authors contributed to the conceptualization and writing of the paper.

Acknowledgements

Lawrence M. Ward was supported by Discovery Grant A9958 from the Natural Sciences and Engineering Research Council (NSERC) of Canada.

Author details

[1] Department of Mathematics, University of British Columbia, Vancouver, BC, Canada.
[2] Department of Psychology and Brain Research Centre, University of British Columbia, 2136 West Mall, Vancouver, BC, V6T 1Z4 Canada.

References

1. McDonnell, M.D., Ward, L.M.: The benefits of noise in neural systems: bridging theory and experiment. Nat. Rev. Neurosci. **12**, 415–425 (2011)
2. Nykamp, D.G., Tranchina, D.: A population density approach that facilitates large-scale modeling of neural networks: analysis and application to orientation tuning. J. Comput. Neurosci. **8**, 19–50 (2000)
3. Greenwood, P.E., McDonnell, M.D., Ward, L.M.: Dynamics of gamma bursts in local field potentials. Neural Comput. **27**, 74–103 (2015)
4. Brunel, N., Hakim, V.: Fast global oscillations in networks of integrate-and-fire neurons with low firing rates. Neural Comput. **11**, 1621–1671 (1999)
5. Longtin, A.: Neuronal noise. Scholarpedia **8**(9), 1618 (2013)
6. Chow, C.C., White, J.A.: Spontaneous action potentials due to channel fluctuations. Biophys. J. **71**, 3013–3021 (1996)
7. White, J.A., Rubinstein, J.T., Kay, A.R.: Channel noise in neurons. Trends Neurosci. **23**, 131–137 (2000)
8. Rowat, P.F., Greenwood, P.E.: The ISI distribution of the Hodgkin-Huxley neuron. Front. Comput. Neurosci. **8**, 111 (2014)
9. Glass, L., Mackey, M.C.: The Rhythms of Life. Princeton University Press, Princeton (1988)
10. Van Vreswijk, C., Sompolinsky, H.: Chaos in neuronal networks with balanced excitatory and inhibitory activity. Science **274**, 1724–1726 (1996)
11. Cutler, C.D.: A theory of correlation dimension for stationary time series. Philos. Trans. R. Soc. B **348**, 348–355 (1994)
12. Greenwood, P.E., Ward, L.M., Wefelmeyer, W.: Statistical analysis of stochastic resonance in a simple setting. Phys. Rev. E **60**(4), 4687–4695 (1999)
13. Stemmler, M.: A single spike suffices. Netw. Comput. Neural Syst. **7**, 687 (1996)
14. Longtin, A.: Stochastic resonance in neuron models. J. Stat. Phys. **70**(1/2), 309–327 (1993)
15. McCulloch, W.S., Pitts, W.: A logical calculus of the ideas immanent in neurons activity. Bull. Math. Biophys. **5**, 115–133 (1943)
16. Bodova, K., Paydarfar, D., Forger, B.: Characterizing spiking in noisy type II neurons. J. Theor. Biol. **365**, 40–54 (2015)
17. Sacerdote, L., Giraudo, M.T.: Stochastic integrate and fire models: a review of mathematical methods and their applications. In: Stochastic Biomathematical Models. Lecture Notes in Mathematics, pp. 99–142. Springer, New York (2013)
18. Kurtz, T.: Strong approximation theorems for density dependent Markov chains. Stoch. Process. Appl. **6**, 223–240 (1978)

© Springer International Publishing Switzerland 2016

P.E. Greenwood, L.M. Ward, *Stochastic Neuron Models*, Mathematical Biosciences Institute Lecture Series, DOI 10.1007/978-3-319-26911-5

69

19. Burkill, A.N.: A review of the integrate-and-fire neuron model: I. homogeneous synaptic input. Biol. Cybern. **95**, 1–19 (2006)
20. Izhikevich, E.M.: Simple model of spiking neurons. IEEE Trans. Neural Netw. **14**, 1569–1572 (2003)
21. Izhikevich, E.M., Edelman, G.M.: Large-scale models of mammalian thalamocortical systems. Proc. Natl. Acad. Sci. U. S. A. **105**(9), 3593–3598 (2008)
22. Hodgkin, A.L., Huxley, A.F.: A quantitative description of membrane current and its application to conduction and excitation in nerve. J. Physiol. **117**, 500–544 (1952)
23. Gillespie, D.T.: Exact stochastic simulation of coupled chemical reactions. J. Phys. Chem. **81**, 2340–2361 (1977)
24. Skaugen, E., Walløe, L.: Firing behaviour in a stochastic nerve membrane model based upon the Hodgkin-Huxley equations. Acta Physiol. Scand. **107**, 343–363 (1979)
25. Rowat, P.F.: Interspike interval statistics in the stochastic Hodgkin-Huxley model: coexistence of gamma frequency bursts and highly irregular firing. Neural Comput. **19**, 1215–1250 (2007)
26. Rowat, P.F., Greenwood, P.E.: Identification and continuity of the distributions of burst-length and interspike intervals in the stochastic Morris-Lecar neuron. Neural Comput. **23**(12), 3094–3124 (2011)
27. Morris, C., Lecar, H.: Voltage oscillations in the barnacle giant muscle fiber. Biophys. J. **35**(1), 193–213 (1981)
28. Ditlevsen, S., Greenwood, P.E.: The Morris-Lecar neuron model embeds a leaky integrate-and-fire model. J. Math. Biol. **67**, 239–259 (2013)
29. Bressloff, P.C.: Metastable states and quasicycles in a stochastic Wilson-Cowan model of neuronal population dynamics. Phys. Rev. E **82**, 051903 (2010)
30. Baxendale, P.H., Greenwood, P.E.: Sustained oscillations for density dependent Markov processes. J. Math. Biol. **63**, 433–457 (2011)
31. Gardiner, C.W.: Handbook of Stochastic Methods for Physics, Chemistry and the Natural Sciences, 2nd edn. Springer, Berlin (1990)
32. FitzHugh, R.: Impulses and physiological states in theoretical models of nerve membrane. Biophys. J. **1**, 445–466 (1961)
33. Nagumo, J., Arimoto, S., Yoshizawa, S.: An active pulse transmission line simulating nerve axon. Proc. IRE **50**, 2061–2070 (1962)
34. van der Pol, B., van der Mark, J.: The heartbeat considered as a relaxation oscillation, and an electrical model of the heart. Lond. Edinb. Dublin Philos. Mag. J. Sci. Ser. 7 **6**(38), 763–775 (1928)
35. Berglund, N., Landon, D.: Mixed-mode oscillations and interspike interval statistics in the stochastic Fitzhugh–Nagumo model. Nonlinearity **8**, 2308–2335 (2012)
36. Lindner, B., Shimansky-Geier, L.: Analytical approach to the stochastic FitzHugh-Nagumo system and coherence resonance. Phys. Rev. E **60**(6), 7270–7276 (1999)
37. Swindale, N.: The development of topography in the visual cortex: a review of models. Netw. Comput. Neural Syst. **7**, 161–247 (1996)
38. Kandel, E.R., Schwartz, J.H., Jessell, T.M.: Principles of Neural Science, 5th edn. McGraw-Hill, New York (2012)
39. Roehrig, C., Rankin, C.H.: Dymods: a framework for modularizing dynamical neuronal structures. Neurocomputing **26–27**, 831–836 (1999)
40. Wilson, H.R., Cowan, J.D.: Excitatory and inhibitory interactions in localized populations of model neurons. Biophys. J. **12**, 1–24 (1972)
41. Kang, K., Shelley, M., Henrie, J.A., Shapley, R.: LFP spectral peaks in V1 cortex: network resonance and cortico-cortical feedback. J. Comput. Neurosci. **29**, 495–507 (2010)
42. Ray, S., Maunsell, J.H.R.: Different origins of gamma rhythm and high-gamma activity in macaque visual cortex. PLOS Biol. **9**(4), 1000610 (2011)
43. McKane, A.J., Biancalani, T., Rogers, T.: Stochastic pattern formation and spontaneous polarisation: the linear noise approximation and beyond. Bull. Math. Biol. **76**, 895–921 (2014)
44. Wallace, E., Benayoun, M., van Dronglen, W., Cowan, J.D.: Emergent oscillations in networks of stochastic spiking neurons. PLoS One **6**(5), 14804 (2011)

45. Borodin, A.N., Salminen, P.: Handbook of Brownian Motion - Facts and Formulae. Probability and Its Applications, 2nd edn. Birkhauser, Basel (2002)
46. Vakeroudis, S.: On the windings of complex-valued Ornstein-Uhlenbeck processes driven by a Brownian motion and by a stable process. arXiv:1209.4027v1 (2012)
47. Lugo, C.A., McKane, A.J.: Quasicycles in a spatial predator-prey model. Phys. Rev. E **78**, 051911 (2008)
48. Asllani, M., Biancalini, T., Fanelli, D., McKane, A.J.: The linear noise approximation for reaction-diffusion systems on networks. arXiv:1305.7318v1 (2013)
49. Van Kampen, N.G.: Stochastic Processes in Physics and Chemistry. North Holland, Amsterdam (1992)
50. Doiron, B., Lindner, B., Longtin, A., Maler, L., Bastian, J.: Oscillatory activity in electrosensory neurons increases with spatial correlation of the stochastic input stimulus. Phys. Rev. Lett. **93**(4), 048101 (2004)
51. Lindner, B., Doiron, B., Longtin, A.: Theory of oscillatory firing induced by spatially correlated noise and delayed inhibitory feedback. Phys. Rev. E **72**, 061919 (2005)
52. Dumont, G., Northoff, G., Longtin, A.: Linear noise approximation for oscillations in a stochastic inhibitory network with delay. Phys. Rev. E **90**, 012702 (2014)
53. Klosek, M.M., Kuske, R.: Multiscale analysis of stochastic differential equations. SIAM Multiscale Model. Simul. **3**, 706–729 (2005)
54. Kuramoto, Y.: Chemical Oscillations, Waves, and Turbulence. Springer, New York (1984)
55. Acebron, J.A., Bonilla, L.L., Vicente, C.J.P., Ritort, F., Spigler, R.: The Kuramoto model: a simple paradigm for synchronization phenomena. Rev. Mod. Phys. **77**, 137–185 (2005)
56. Strogatz, S.H.: From Kuramoto to Crawford: exploring the onset of synchronization in populations of coupled oscillators. Phys. D **143**, 1–20 (2000)
57. Greenwood, P.E., McDonnell, M.D., Ward, L.M.: A Kuramoto coupling of quasi-cycle oscillators. arXiv:1511.04124v2
58. DeVille, R.E.L., Peskin, C.S., Spencer, J.H.: Dynamics of stochastic neural networks and the connection to random graph theory. Math. Model. Nat. Phenom. **5**(2), 26–66 (2010)
59. Turing, A.M.: The chemical basis of morphogenesis. Philos. Trans. R. Soc. B **237**, 37–72 (1952)
60. Butler, T., Goldenfeld, N.: Fluctuation-driven Turing patterns. Phys. Rev. E **84**(011112) (2011)
61. Biancalani, T., Fanelli, D., Di Patti, F.: Stochastic Turing patterns in the Brusselator model. Phys. Rev. E **81**, 046215 (2010)
62. Butler, T.C., Benayoun, M., Wallace, E., van Drongelen, W., Goldenfeld, N., Cowan, J.: Evolutionary constraints on visual cortex architecture from the dynamics of hallucinations. Proc. Natl. Acad. Sci. U.S.A. **109**(2), 606–609 (2012)
63. Bressloff, P.C., Cowan, J.D., Golubitsky, M., Thomas, P.J., Wiener, M.C.: Geometric visual hallucinations, Euclidean symmetry and the functional architecture of the striate cortex. Philos. Trans. Biol. Sci. **356**, 299–330 (2001)
64. Ermentrout, B., Cowan, J.D.: A mathematical theory of visual hallucination patterns. Biological Cybernetics **34**, 137–150 (1979)
65. Neiman, A., Shimansky-Geier, L., Cornell-Bell, A., Moss, F.: Noise-enhanced phase synchronization in excitable media. Phys. Rev. Lett. **83**, 4893–4897 (1999)
66. Biancalani, T., Galla, T., McKane, A.J.: Stochastic waves in a Brusselator model with nonlocal interaction. Phys. Rev. E **84**, 026201 (2011)
67. Marr, D.: Vision: A Computational Investigation Into the Human Representation and Processing of Visual Information. Freeman, New York (1982)
68. Siebert, J., Schöll, E.: Front and Turing patterns induced by Mexican-hat-like nonlocal feedback. Europhys. Lett. **109**, 40014 (2015)
69. Milo, R., ShenÄêOrr, S., Itzkovitz, S., Kashtan, N., Chklovskii, D., Alon, U.: Network motifs: simple building blocks of complex networks. Science **298**, 824–827 (2002)
70. McDonnell, M.D., Mohan, A., Stricker, C., Ward, L.M.: Input-rate modulation of gamma oscillations is sensitive to network topology, delays and short-term plasticity. Brain Res. **1434**, 162–177 (2012)

71. McDonnell, M.D., Yaveroglu, O.N., Schmerl, B., Iannella, N., Ward, L.M.: Motif-role-fingerprints: the building-blocks of motifs, clustering-coefficients and transitivities in directed networks. PLOS ONE **9**(12), 114503 (2014)
72. Silvestri, G., Ward, L.M.: Local connectivity does not determine global synchronization in spiking neural networks. Unpublished ms, UBC (2014)
73. Tarjan, R.: Depth-first search and linear graph algorithm. SIAM J. Comput. **1**, 146–160 (1972)
74. Ward, L.M.: Physics of neural synchronisation mediated by stochastic resonance. Contemp. Phys. **50**(5), 563–574 (2009)
75. Kozma, R., Puljic, M.: Random graph theory and neuropercolation for modeling brain oscillations at criticality. Curr. Opin. Neurobiol. **31**, 181–188 (2015)
76. Kozma, R., Puljic, M., Balister, P., Bollobas, B., Freeman, W.J.: Phase transitions in the neuropercolation model of neural populations with mixed local and non-local interactions. Biol. Cybern. **92**, 367–379 (2005)
77. McDonnell, M.D., Ward, L.M.: Small modifications to network topology can induce stochastic bistable spiking dynamics in a balanced cortical model. PLOS ONE **9**(4), 88254 (2014)
78. Litwin-Kumar, A., Doiron, B.: Slow dynamics and high variability in balanced networks with clustered connections. Nat. Neurosci. **15**, 1498–1505 (2012)
79. van der Hofstad, R., Holmes, M., Kuznetsov, A., Ruszel, W.: Strongly reinforced Polya urns with graph-based competition. arXiv:1406.0449v1 (2014)
80. Nunez, P.L.: The brain wave equation: a model for the EEG. Math. Biosci. **21**, 279–297 (1974)
81. Nunez, P.L.: Toward a quantitative description of large-scale neocortical dynamic function and EEG. Behav. Brain Sci. **23**, 371–437 (2000)
82. Nunez, P.L., Srinivasan, R.: Neocortical dynamics due to axon propagation delays in cortico-cortical fibers: EEG traveling and standing waves with implications for top-down influences on local networks and white matter disease. Brain Res. **1542**, 138–166 (2014)
83. Wingeier, B.M.: A high resolution study of coherence and spatial spectra in human EEG. Ph.D. thesis, Tulane University (2004)
84. Deco, G., Jirsa, V.K., McIntosh, A.R.: Resting brains never rest: computational insights into potential cognitive architectures. Trends Neurosci. **36**, 268–274 (2013)
85. Jirsa, V.K.: Neural field dynamics with local and global connectivity and time delay. Philos. Trans. R. Soc. A **367**, 1131–1143 (2009)
86. Jirsa, V.K., Haken, H.: Field theory of electromagnetic brain activity. Phys. Rev. Lett. **77**, 960–963 (1996)
87. Bojak, I., Oostendorp, T.F., Reid, A., Kotter, R.: Connecting mean field models of neural activity to EEG and fMRI data. Brain Topogr. **23**, 139–149 (2010)
88. Amunts, K., Lepage, C., Borgeat, L., Mohlberg, H., Dickscheid, T., et al.: Bigbrain: an ultrahigh-resolution 3D human brain model. Science **340**, 1472–1475 (2013)

Index

A
Action potential, 2, 22, 27, 29, 34
Action potential density, 64
Alpha
  frequency, 65
  rhythm, 59, 65
Asymptotic, 2, 29
Axon, 2, 43, 64

B
Bifurcation diagram, 17, 19, 20, 26
BigBrain, 66
Binary neuron, 2, 9–11, 34, 43, 61
Bistable, 5, 17
Blood-oxygenation-level-dependent (BOLD), 66
Brownian motion, 5, 12, 14, 25, 35, 37–39, 44
Brunel, N., 3, 42, 45
Brusselator, 55, 57

C
Connectivity, 47, 60, 64, 66
Coupling, 3, 45, 47, 50, 54, 58, 62

D
Damped oscillation, 23, 39, 44, 64
Delay, 3, 34, 42–45, 65, 66
Density-dependent Markov process, 39, 52
Diffusion, 3, 12, 18, 23, 36, 39, 49–51, 53–56, 61, 62, 65

E
Edelman, G.M., 65
Eigenvalue, 23, 26, 37, 42, 44, 45, 54, 56
Excitation, 34, 50, 51
Excitation-inhibition, 2, 3, 11, 34, 35, 38, 58, 60
Excitatory, 2, 3, 9, 11, 34, 35, 38, 39, 49, 58–60, 64

F
Firing, 2–5, 9, 10, 12–14, 17, 20–23, 25–30, 33–35, 38, 42, 43, 46, 49, 54, 58, 60, 62
FitzHugh-Nagumo, 29, 54
Fixed point, 7, 17, 20, 21, 23–26, 28, 29, 35, 36, 39–41, 56
Focus, 12, 19, 40, 43, 49
Fokker-Planck, 12, 42
Fourier transform, 40, 42, 55, 56
Functional magnetic resonance imaging (fMRI), 66

G
Gamma
  band, 59
  frequency, 36, 38
  rhythm, 59
Graph-based model, 61–62

H
Hakim, V., 3, 42, 45
Hodgkin-Huxley, 5, 7, 14–18, 27
Human brain atlas, 66

© Springer International Publishing Switzerland 2016
P.E. Greenwood, L.M. Ward, *Stochastic Neuron Models*, Mathematical Biosciences
Institute Lecture Series, DOI 10.1007/978-3-319-26911-5

Printed in the United States
By Bookmasters